Regional Problems of Earth Remote Sensing (RPERS 2021)

E3S Web of Conferences Volume 333 (2021)

Krasnoyarsk, Russia
14 - 17 September 2021

Editors:

Y.A. Maglinets
M.V. Noskov
G.M. Tsibul'skii

ISBN: 978-1-7138-4178-4

Printed from e-media with permission by:

Curran Associates, Inc.
57 Morehouse Lane
Red Hook, NY 12571

proceedings.com

Some format issues inherent in the e-media version may also appear in this print version.

For additional information, please contact EDP Sciences – Web of Conferences at the address below.

EDP Sciences – Web of Conferences
17, Avenue du Hoggar
Parc d'Activité de Courtabœuf
BP 112
F-91944 Les Ulis Cedex A
France

Phone: +33 (0) 1 69 18 75 75
Fax: +33 (0) 1 69 28 84 91

contact-edps@webofconferences.org

Additional copies of this publication are available from:

Curran Associates, Inc.
57 Morehouse Lane
Red Hook, NY 12571 USA
Phone: 845-758-0400
Fax: 845-758-2633
Email: curran@proceedings.com
Web: www.proceedings.com

TABLE OF CONTENTS

edp sciences

Web of Conferences

Statement of Peer review

In submitting conference proceedings to *Web of Conferences*, the editors of the proceedings certify to the Publisher that

1. They adhere to its **Policy on Publishing Integrity** in order to safeguard good scientific practice in publishing.
2. All articles have been subjected to peer review administered by the proceedings editors.
3. Reviews have been conducted by expert referees, who have been requested to provide unbiased and constructive comments aimed, whenever possible, at improving the work.
4. Proceedings editors have taken all reasonable steps to ensure the quality of the materials they publish and their decision to accept or reject a paper for publication has been based only on the merits of the work and the relevance to the journal.

Title, date and place of the conference

Regional Problems of Earth Remote Sensing RPERS 2021, September 14- September 17, 2021

Proceedings editor(s):

Y.A. Maglinets

Date and editor's signature

Statement of Peer review

In submitting conference proceedings to *Web of Conferences*, the editors of the proceedings certify to the Publisher that

1. They adhere to its **Policy on Publishing Integrity** in order to safeguard good scientific practice in publishing.
2. All articles have been subjected to peer review administered by the proceedings editors.
3. Reviews have been conducted by expert referees, who have been requested to provide unbiased and constructive comments aimed, whenever possible, at improving the work.
4. Proceedings editors have taken all reasonable steps to ensure the quality of the materials they publish and their decision to accept or reject a paper for publication has been based only on the merits of the work and the relevance to the journal.

Title, date and place of the conference

Regional Problems of Earth Remote Sensing RPERS 2021, September 14- September 17, 2021

Proceedings editor(s):

M.V. Noskov

Date and editor's signature

edp sciences

Web of Conferences

Statement of Peer review

In submitting conference proceedings to *Web of Conferences*, the editors of the proceedings certify to the Publisher that

1. They adhere to its **Policy on Publishing Integrity** in order to safeguard good scientific practice in publishing.
2. All articles have been subjected to peer review administered by the proceedings editors.
3. Reviews have been conducted by expert referees, who have been requested to provide unbiased and constructive comments aimed, whenever possible, at improving the work.
4. Proceedings editors have taken all reasonable steps to ensure the quality of the materials they publish and their decision to accept or reject a paper for publication has been based only on the merits of the work and the relevance to the journal.

Title, date and place of the conference

Regional Problems of Earth Remote Sensing RPERS 2021, September 14- September 17, 2021

Proceedings editor(s):

G.M. Tsibul'skii

Date and editor's signature

17, avenue du Hoggar - PA de Courtabœuf - BP 112 - 91944 Les Ulis Cedex A (France)
Tél. : 33 (0)1 69 18 75 75 - Fax : 33(0)1 69 07 45 17 - www.edpsciences.org

CROP YIELD MAP DEVELOPMENT BASED ON SATELLITE AND UAV DATA

Irina Botvich[1,2*], *Anatoly* Shevyrnogov[1], and *Dmitriy* Emelyanov[1]

[1]Institute of Biophysics SB RAS,660036 Krasnoyarsk, Russia
[2]Siberian Federal University, 660041 Krasnoyarsk, Russia

Abstract. Agriculture is one of the promising areas for the introduction of remote sensing technologies. With its help, you can timely receive a wide range of dynamic information about the conditions for the crops growth and development and make the necessary adjustments to obtain the planned result. The paper presents the results of early forecasting of spring wheat yields based on Dove (PlanetScope) satellite data from Planet Labs and DJI P4 Multispectral unmanned data. The maps of the yield of spring wheat were constructed using satellite and unmanned data with a spatial resolution of 3 m and 5 cm, respectively. A statistical assessment of the inhomogeneity of the spectral optical characteristics of agricultural crops has been carried out. The degree of correlation dependence between the value of the integral of the index (NDVI, VARI, MSAVI2, ClGreen) and the yield in different periods of the growing season has been determined.

Keywords: precision farming, yield, growing season, spring wheat, PlanetScope, DJI P4 Multispectral

1. Introduction

The widespread introduction of energy-saving technologies for the cultivation of agricultural crops is aimed at reducing material and technical costs in the crop production. An increase in the level of profitability of agricultural production in modern conditions is possible only on the basis of the introduction of modern energy-saving technologies for growing plants using elements of precision farming Remote sensing methods make it possible to predict the yield of field crops and, at the same time, to assess the crops state (weediness, plant density, damage by pests and diseases, the state of crops after overwintering, etc.) and, ultimately, to manage production process. The purpose of this study is to develop a precision farming method for early forecasting of grain yield based on remote sensing data (satellite and unmanned aerial vehicle - UAV) and geobotanical studies.

* Corresponding author: irina.pugacheva@mail.ru

2. Objects and Methods

The studies were carried out on the lands of the Minderlinskoye educational farm in the Sukhobuzimsky district of the Krasnoyarsk Territory during the growing seasons of 2019-2020. Figure 1 shows the location of the experimental field and its structure. The experimental field consists of 5 strips. Their average length is 650 meters. The direction of the strips is from west to east, the width of each strip is 20 m. Fertilization was carried out along the northern part of each strip (10 m wide), the southern part remained without fertilizers. Each strip is sown with a specific type of crops.

The paper presents an analysis of the data processing results on the strips sown with wheat "Novosibirskaya-15". The fields are divided into 4 test areas in accordance with four types of soil cultivation: plowing (20-22 cm) - a; sweep treatment (20-22 cm) -b; surface treatment (discator by 8-10 cm) - c; no tillage - d. For the purity of the experiment, all test areas are separated from each other by protective stripes 5 meters wide. Accounting for the yield of wheat grain was carried out by a continuous method using a combine harvester in 6 replicates in 2019 and 4 replicates in 2020, for each type of soil treatment. The grain yield was calculated in centners per hectare, taking into account the conversion to 14% moisture and 100% purity.

Fig. 1. Study site and its structure by types of processing and crops. Lines 3 and 5 are sown with spring wheat. Borders of different types of tillage: d, c, b, a - in the direction from left to right.

The study is based on satellite data from PlanetScope (spatial resolution 3 m) and UAV data from the DJI P4 Multispectral (spatial resolution 5 cm). The spectral indices NDVI (Normalized Difference Vegetation Index) [1], VARI (Visible Atmospherically Resistant Index) [2], MSAVI2 (Modified Soil Adjusted Vegetation Index) [3], ClGreen (Green chlorophyll index) were calculated using satellite and unmanned spectrophotometric data. [4] (Table 1).

Table 1. Satellite and unmanned spectral channels for calculating indices:
b1 – 1 channel reflectance (455 -515 nm) of PlanetScope; 450 nm (±16 nm) DJI P4 Multispectral;
b2 – 2 channel reflectance (500 – 590 nm) of PlanetScope; 560 nm (±16 nm) DJI P4 Multispectral;
b3 – 3 channel reflectance (590 – 670 nm) of PlanetScope; 650 nm (±16 nm) DJI P4 Multispectral;

b4 – 4 channel reflectance (780 - 860 nm) of PlanetScope; 5 channel 840 nm (±16 nm) DJI P4 Multispectral.

Index	Formula
NDVI	(b4-b3)/(b4+b3)
VARI	(b2-b3)/(b2+b3-b1)
MSAVI2	$(2*b4+1-((2*b4+1)^2-8*(b4-b3))^{1/2})/2$
ClGreen	(b4/b2)-1

The calculation of the curvilinear integral of the studied indices (I) is carried out according to the formula:

$$I = \sum_{k=1}^{n}\left(\left(\frac{V_k + V_{k+1}}{2}\right)\cdot\left(d_{k+1} - d_k\right)\right) \quad (1)$$

where V_k is the value of the vegetation index, d_k is the time (day of survey), n is the number of measurements.

3. Results and discussion

To determine the relationship between the integrals of the index values and the yield of wheat, the coefficient of determination (r^2) between these variables was calculated. Wheat yield was measured in 4 replicates for each type of treatments (64 plots in total). For the same areas, the values of the integrals of the indices NDVI, MSAVI2, VARI, ClGreen were determined. Analysis of the relationship between the yield and the value of the integral during the growing season of 2020 (according to satellite data) showed the presence of a more pronounced plateau. At the end of June (06/26/2020), the r^2 value reaches high values, more than 0.7 for MSAVI2 and ClGreen. For NDVI values, r^2 are significantly lower than 0.59. The lowest values are obtained for the VARI 0.53 integral (fig. 2).

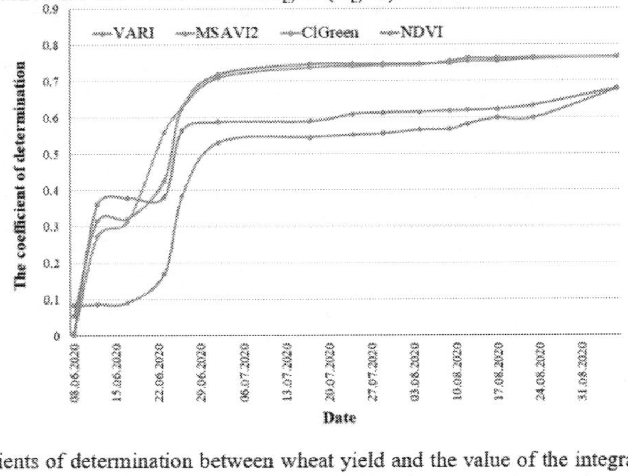

Fig. 2. Coefficients of determination between wheat yield and the value of the integral of the NDVI, MSAVI2, VARI, ClGreen indices, calculated using satellite data during the growing season of 2020.

Analysis of the relationship between the yield and the integral value during the growing season of 2020 (according to unmanned data) showed that from the beginning of July the value of r^2 reaches sufficiently significant values (from 0.42 to 0.5) for all indices, while the

NDVI integral has the minimum value, the maximum ClGreen and MSAVI2. The maximum values were reached by August 20 - 0.64 (ClGreen) and 0.69 MSAVI2. The maximum values of the NDVI integral are 0.71 by the first days of September.

Analysis of the relationship between the yield and the integral value during the growing season of 2019 (according to satellite data) showed that the coefficient of determination r^2 has the highest values when using the NDVI integral. Since the first days of July, r^2 has exceeded 0.86. The VARI integral had the lowest values during this time period. At the end of the growing season (before harvesting), all the integrals of the indices showed a high degree of relationship with yield - from 0.84 for VARI to 0.93 for NDVI, the determination values MSAVI2, ClGreen are 0.9.

Thus, an analysis of the data obtained over two years of research on satellite and unmanned measurements allows us to conclude that the most optimal and reliable is the use of integrals of the MSAVI2 index.

Representation of information in the form of accumulated values of integrals MSAVI2 shows that, in comparison with the spatial variation of direct values of MSAVI2, a structure that is stable in time appears for all types of treatments. At the same time, there is a clear dependence of the change in the spatial structure of the MSAVI2 integrals in time (Figs. 3, 4).

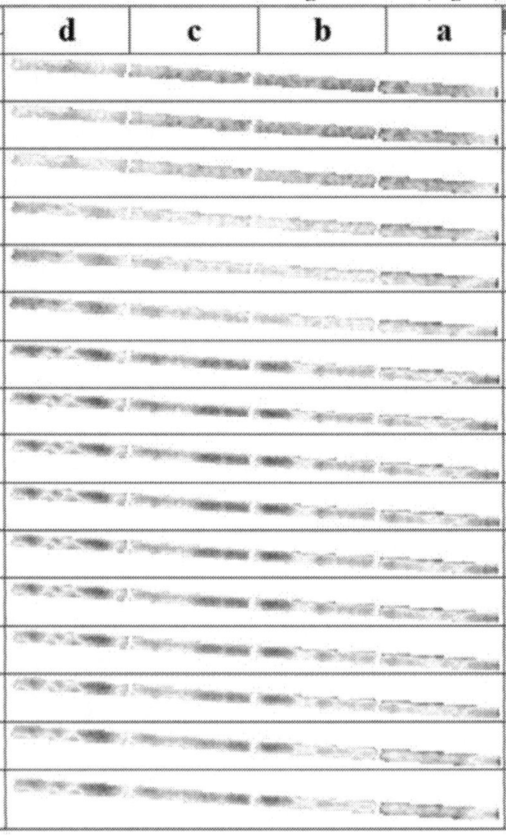

Date: beginning	Date: end	min	max
23.05.2020	08.06.2020	5.88	8.91
23.05.2020	12.06.2020	7.81	11.32
23.05.2020	17.06.2020	10.65	14.66
23.05.2020	23.06.2020	14.15	18.66
23.05.2020	26.06.2020	15.98	20.59
23.05.2020	02.07.2020	20.14	24.83
23.05.2020	17.07.2020	30.05	35.08
23.05.2020	24.07.2020	34.29	39.69
23.05.2020	29.07.2020	37.00	42.97
23.05.2020	04.08.2020	40.03	46.69
23.05.2020	09.08.2020	42.62	49.70
23.05.2020	12.08.2020	43.93	51.21
23.05.2020	17.08.2020	45.73	53.42
23.05.2020	23.08.2020	48.25	56.39
23.05.2020	06.09.2020	54.32	64.05
23.05.2020	15.09.2020	56.72	67.30

Fig. 3. Cartograms of the MSAVI2 wheat crops integral during the growing season (from May 23 to September 15, 2020) obtained from satellite data.

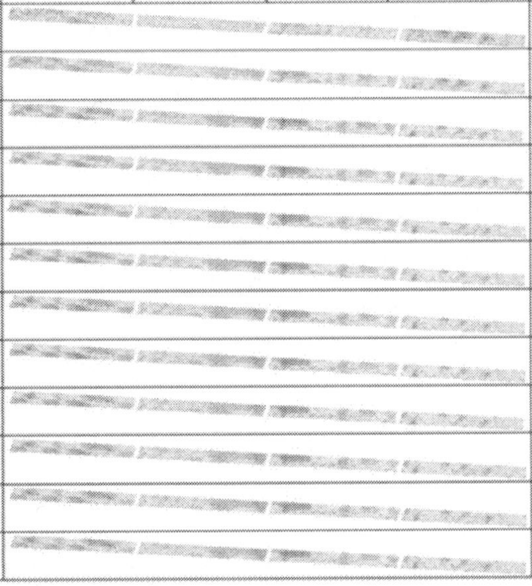

Date		d	c	b	a		
beginning	end					min	max
18.06.2020	26.06.2020					0.012	0.103
18.06.2020	03.07.2020					0.029	0.198
18.06.2020	13.07.2020					0.058	0.336
18.06.2020	22.07.2020					0.079	0.464
18.06.2020	30.07.2020					0.096	0.543
18.06.2020	06.08.2020					0.107	0.581
18.06.2020	13.08.2020					0.121	0.606
18.06.2020	20.08.2020					0.133	0.624
18.06.2020	27.08.2020					0.140	0.634
18.06.2020	03.09.2020					0.147	0.643
18.06.2020	10.09.2020					0.153	0.633
18.06.2020	16.09.2020					0.158	0.661

Fig. 4. Cartograms of the MSAVI2 wheat crops integral during the growing season (from June 18 to September 16, 2020), obtained from unmanned data.

It was found that the spatial-temporal heterogeneity of satellite data is smoothed out on the cartograms of the vegetation indices integrals, as well as on unmanned data, but at the same time a fine spatial structure is preserved on unmanned data. These results allow to conclude that for early forecasting without taking into account the fine spatial structure, it is more preferable to use satellite data, and when assessing the intra-field structure of yield, less reliable results on specific yield can be obtained, but more accurate in the location of plots with higher and lower productivity.

To predict the wheat yield at the beginning of July 2020, linear regression models were calculated, in which the integral values under the MSAVI2 curve at different time periods were used as parameters. The multiple linear model for forecasting wheat from satellite data has the form for 6 variables (determination coefficient 0.72):

$$Y=-50.73+14.17\times Imsavi2_1+5.38\times Imsavi2_2-13.18\times Imsavi2_3-5.22\times Imsavi2_4-19.14\times Imsavi2_5+23.4\times Imsavi2_6 \quad (2)$$

A multiple linear model for forecasting wheat based on unmanned data has the form for 2 variables (coefficient of determination 0.49):

$$Y=14.33-627.23\times Imsavi2_1+429.31\times Imsavi2_2 \quad (3)$$

where $Imsavi2_i$ is the integral value under the MSAVI2 curve at time i; Y is the yield value, c / ha.

The calculations made it possible to construct a yield cartogram of the study site. Figures 5 - 6 (I and II) show the spatial distributions of the study site wheat yield, obtained from the model calculations results using unmanned (I) and satellite data (II).

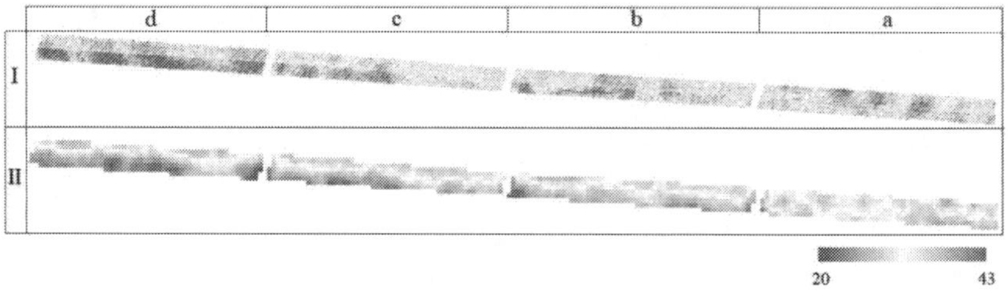

Fig. 5. Maps of the spring wheat yield by fallow (3rd strip), obtained from the results of a model calculation using unmanned data (I) and the results of a model calculation using satellite data (II).

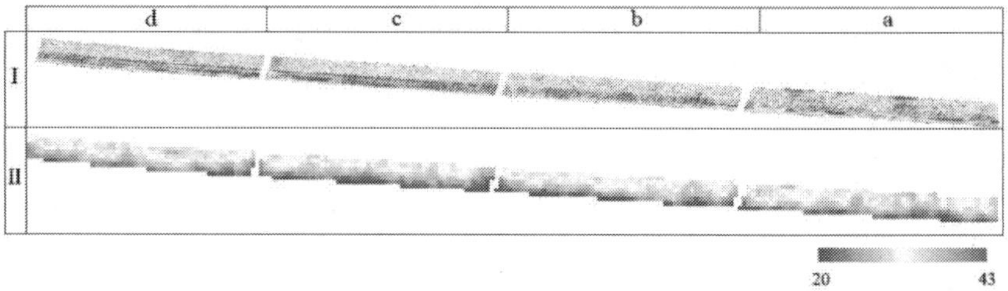

Fig. 6. Maps of the spring wheat yield by corn (5th strip), obtained from the results of a model calculation using unmanned data (I) and the results of a model calculation using satellite data (II).

Comparison of cartograms obtained from the results of model calculations using unmanned (I) and satellite (II) data (Fig. 5, 6) show the similarity of the spatial distribution of yield. The higher spatial resolution of unmanned data (5 cm) gives a clearer picture of crop heterogeneity, including the presence and absence of fertilized background. In the future, the maps obtained from unmanned data can be successfully applied for differentiated fertilization and selection of soil samples.

The study was carried out with the support of the Krasnoyarsk Regional Science Foundation within the framework of the project: "Development and implementation of a method for early forecasting of grain yield in the Siberian region according to remote sensing data of the Earth".

References

1. C.F. Jordan, Ecology, Ecology, **50** (1969)
2. A.A. Gitelson, R. Stark, U.Grits, D.Rundquist, Y.Kaufman, D. Derry, International Journal of Remote Sensing, **23** (2002)
3. J. Qi, A. Chehbouni, A.R. Huete, Y.H. Kerr, S. Sorooshian, Remote Sensing of Environment, **48** (1994)
4. G. Dall'Olmo, A.A.Gitelson, D.C. Rundquista, B. Leavitt, T. Barrow, J.C. Holz, Remote Sensing of Environment, **96** (2005)

Visual Study of Multi-Dimensional Data in Remote Earth Sensing Problems

Boris Dobronets[1,*], *Olga* Popova[1], and *Alexey* Merko[1]

[1]Institute of Space and Information Technology, Siberian Federal University, Kirenskogo 26, Krasnoyarsk, 660074 Russia

Abstract. The article deals with the problem of multidimensional data presentation based on the application of the scientific visualization method. A method of visual-interactive animation based on dynamic representation of images in the form of 2D cross-sectional models is proposed. The effect of multidimensionality is achieved through the choice of 2D models, the variability of the directions of dynamic sections and the use of stereoscopic vision capabilities.

The task of extracting meaningful information from multidimensional data obtained using Earth remote sensing (ERS) is usually very difficult. Currently, a lot of work is being done on the use of Earth remote sensing data for solving many problems of the national economy, where it is important, first of all, to analyze the spatio-temporal changes in natural and man-made processes. Space monitoring makes it possible to obtain uniform and comparable in quality objective information of huge volumes simultaneously over vast territories, which is unattainable for any ground survey. An important feature of the incoming information is its large volume, which complicates its processing and interpretation.

It should be noted that understanding of the higher dimensions is limited. The visualization is mainly presented in three dimensions, including additional animation. Recently, approaches have been actively used for the visual presentation of multidimensional information using projections or abstractions, namely, from dimensionality reduction to visual coding and from quantitative analysis to interactive modeling.

Various techniques are currently used to visually explore multidimensional data. Among them, we note the use of perspective, the removal of hidden lines and surfaces, the use of stereographic images to represent the image of an object that deceives our visual system in the perception of space, or a three-dimensional image drawn on a clearly flat computer screen. When analyzing images in a virtual reality headset, motion parallax is applied, which helps to perceive depth, even if the display in front of us is really nothing more than an array of colored dots [1–4].

Many studies of multivariate data today use the concept of scientific visualization. Scientific visualization is based on three stages of information transformation: data transformation, visual display, viewing the transformed data. Data transformation follows analytical techniques such as dimensionality reduction, regression, clustering, feature extraction, topological analysis, data sampling, and abstraction. Visualization, as the key to most visual coding tasks, focuses on organizing information from the stage of converting data to visual presentation. Viewing the converted data follows screen-space and rendering-oriented techniques, and measures screen space to reduce artifacts and highlight important features.

*e-mail: BDobronets@yandex.ru

This study focuses on the development of visual interactive animation techniques for processing, presenting and analyzing multidimensional data. It is proposed to use the method of visual-interactive modeling of spatial data to obtain the maximum possible information about the object under study. For this, an approach based on scientific visualization and visual analytics methods is used. It allows you to organize a human-machine interface that enhances the analytical abilities of a person using methods such as expanding a person's random access memory through the use of special visualization based on placing information in space in accordance with temporal relationships and organizing a controlled environment for a user to work in a space of parametric values. This organization of visual presentation and interfaces provides a person with the ability to immediately see, explore and understand huge amounts of information.

Let's consider the main methods of scientific visualization. First of all, we note the methods for reducing the dimensionality of spatial features and characteristics, which is a key factor for many visualization tasks. The main ones can be distinguished. So linear projection includes many classical methods, such as the principal component analysis (PCA), the multivariate scaling (MDS), the linear discriminate analysis (LDA) and various methods of the factor analysis [5].

Clustering is one of the most widely used methods of multivariate data analysis.Regression analysis of multivariate data is an extensive and active field of research in its own right. Researchers also use regression to summarize data. Methods like this depict data structures through wireframes and use a clustering algorithm followed by the construction of a minimal tree of cluster centroids to determine possible trends in the data. Here, regression is used as a post-processing step in algorithms to represent the totals of the extracted subsets of the data. Design management techniques are also common in which, in addition to a multidimensional input space, the user has one or more output or response variables that he wants to explore, and the results require high-quality expertise, or decision justification. We will focus on the combined application of the visualization method and regression analysis [5].

Note that visualization is primarily aimed at increasing the information content of the person solving the problem. Therefore, it is important to note that for the number of dimensions less than three, the implementation of visual data analysis is not particularly difficult, because a person has two-dimensional vision and geometric images that fit into his consciousness and representations for spaces with the number of dimensions $n \geq 3$. For multidimensional objects with a large number of dimensions, a person does not have such geometric images. Therefore, it is necessary to project into nested spaces with a standard number of dimensions. Apart from the three spatial dimensions, the only other dimension that we meet in the real world is time. Once a certain frame fusion rate is reached, there is a natural tendency to link different static images together (beta motion). Images displayed faster than about 20 Hz will appear continuously without flicker. During animation, processing each frame should also include restoring the render for each new time step, or perhaps reading a data file to get a new time step. One alternative to animation is to tiled the display with different frames. Note that the animation gives an overview of the object, while looking at the frames reveals small details that might otherwise have been overlooked. Therefore, ideally, we aim to provide both mechanisms for displaying multidimensional datasets [6–8].

The main feature of the considered approach to the visual-interactive representation of multidimensional data with the number of dimensions $n \geq 3$ is their dynamic representation in the form of 2D cross-sectional models. The effect of the multidimensionality is achieved through the choice of 2D models, the variability of the directions of dynamic sections and the use of stereoscopic vision capabilities.

To improve the quality of perception, it is necessary to reduce the effect of overlapping sections. One of these methods is the ability to show not all information on the sections, but only a part. For example, apply contours using color, showing information only at the nodes of the grid. Stereo projections are used to improve perception. To reproduce isolines on sections, you can simultaneously reproduce several close sections, varying their brightness and transparency. The parameters of the method of visual-interactive animation are the distance between the planes and the speed of movement of the sections. The thickness and transparency of the isolines smoothly change depending on the plane number. The contour lines are most vividly represented in the foreground (active sections).

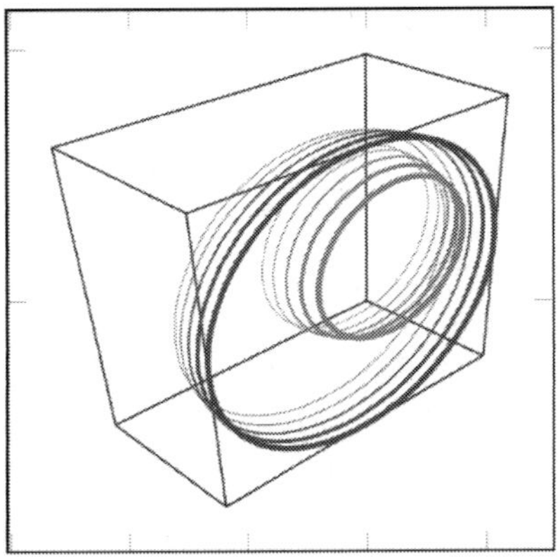

Figure 1. Isolines on sections

Figure 1 shows a fragment of the animation of the function with five sections. These sections are formed by five parallel planes. The parameters of the method of visual-interactive animation are the distance between the planes and the speed of movement of the sections. The thickness and transparency of the isolines smoothly change depending on the plane number. Isolines are most clearly shown in the foreground (active sections). The contours are shown in conventional colors, which corresponds to the value of the function. The red color corresponds to the value of the parameter $t = t_1$, the blue color $t = t_2$.

Let us dwell in more detail on the representation of an object in the form of dynamic sections. The purpose of this approach is to represent the movement of sections as an image of an object. Stereo pairs are generated for better perception. The constructed stereo image can be viewed using various equipment. Figure 2 shows a stereo image of dynamic sections.

Let's consider the most affordable. A stereoscope is an optical instrument with two eyepieces. Typically used for viewing stereo slides, it is easy to put in a laptop or communicator with an oblong high-resolution screen. A stereo display is an optical instrument by which two planar images are combined so that the observer gets a representation of the embossed object. You can use a modern TV that allows you to view stereo images through special glasses. A virtual helmet (VR HMD) is a device that displays separate images for each eye. The result is a stereo effect. The constructed stereo images were tested through a stereoscope by uploading a video file to a smartphone. The results of testing the method of visual-interactive animation showed that the most realistic picture of the presentation of three-dimensional data was obtained with dynamic reproduction of stereo images of sections with contours in conventional colors. Currently, research is underway to implement this approach for objects of dimension more than three.

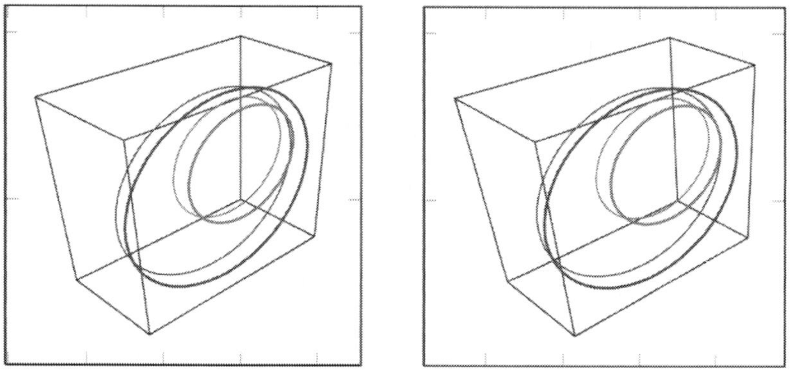

Figure 2. Stereo image of dynamic sections

The presented approach, combined with the use of numerical modeling methods, makes it possible to study remote sensing data in the tasks of studying terrain reliefs, for conducting hydrological surveys, studying agricultural land, and so on. It significantly increases the interpretability and information content of research results for various specialists.

References

[1] A.E. Bondarev, V.A. Galaktionov, V.M. Chechetkin, Computational Mathematics and Mathematical Physics **51**(4), 624–636 (2011)

[2] O.P. Maslennikov, I.E. Milman, A.E. Safiulin, A.E. Bondarev, Sh.U. Nizametdinov, V.V. Pilyugin, Scientific Visualization, **6**, 30–49 (2014)

[3] O.A. Popova, Informatization and communication **5** 93–96 (2019)

[4] H. Wright *Introduction to Scientific Visualization* (London, Springer-Verlag, 2007)

[5] S. Liu, D. Maljovec, B. Wang, P.-T. Bremer, V. Pascucci, IEEE Transactions on Visualization and Computer Graphics. **23**(3), 1249–1268 (2017)

[6] J. Steele (Eds.) *Beautiful Visualization* (OŔeilly Media, Inc., 2010)

[7] Q. Zhang, R. Segall, M. Cao, (Hershey, New York, Information Science reference, 2011)

[8] J. Dill, *Expanding the Frontiers of Visual Analytics and Visualization* (London: Springer-Verlag, 2012)

Detection of fallow lands overgrowth from satellite images based on wavelet analysis

Larisa Evstratova[1*], *Anton* Antoshkin[1]

[1]State University of Land Use Planning, 105064, 15 Kazakova str., Moscow, Russian Federation

Abstract. To improve the efficiency of agricultural lands condition and use monitoring, method for detecting fallow lands overgrowth with tree and shrub vegetation using multi-temporal multispectral satellite images of high spatial resolution is proposed. The task of localization of overgrowth is solved on the basis of multi-temporal satellite images wavelet analysis. This technique allows to operationally and automatically detect areas of fallow lands overgrowth with tree and shrub vegetation. The proposed methodology was tested using satellite images obtained from Quickbird, GeoEye and WorldView satellites. For each image, a fourth-order wavelet decomposition was performed in terms of the Daubeshi wavelet functions. Comparing the different levels of the images wavelet transform, it is possible to divide the areas without tree and shrub vegetation, and where overgrowth was depicted. The method proposed by the authors for detecting changes based on wavelet analysis has shown that this method practically does not depend on the influence of shooting conditions and, accordingly, is more effective at detecting changes.

When solving the problem of effective involvement in the turnover of agricultural land, the question arises of determining the location and area of unused land plots subject to tree and shrub vegetation overgrowth. The most operational and reliable data for solving this problem are multispectral images obtained by space survey systems. However, automating the process of detecting changes based on multi-time satellite images is difficult task, in addition, visual analysis is often used when performing production work, which requires a lot of labor costs.

Thus, the search for new approaches for automated recognition of terrain objects images and identification of changes in multispectral satellite images at different times that would allow for prompt and effective the detecting of unused agricultural land and overgrowth long fallow is an urgent task.

Modern automated methods of decryption make it possible to recognize various objects with sufficient reliability. However, when analyzing multispectral images of different times, the spectral characteristics of the land lots may change, since they significantly depend on the shooting conditions and the object status. Therefore, algorithms in which the main decryption feature is the object spectral brightness cannot serve as a sufficient decryption feature [1; 2]. Many algorithms have been developed for detecting changes from

* Corresponding author: lge_21@mail.ru

satellite images, but there is no universal method, as a rule, the algorithms are adapted for certain type of data (survey systems) and changes categories. In addition, different objects classes when recognized from satellite images often do not mutually exclude each other, which make it difficult to perform Change Detection procedures [1; 3].

One of the solutions to the problem is the use of structural features of multispectral images of reference plots in satellite images. It is known that structural features are more stable, characterize the internal properties of object images directly related to its structure and practically does not change over time [4; 5; 6], which means that algorithms based on these properties are more efficient than using spectral properties of images. Such features are obtained by Fourier transformation, wavelet transformation, by transformation using specially specified basis functions of various types (for example, spherical functions) [7; 8; 9].

The paper presents the results of researches of the proposed technique for the rapid detection of fallow lands overgrowth with woody and shrubby vegetation on multi-time multispectral satellite images based on the wavelet transform.

The purpose of this work is to assess the possibility of using wavelet analysis when processing multi-time images to localize the overgrowth of agricultural land plots with tree and shrub.

The paper provides an example of using wavelet functions in which decomposition is performed for different scales of the functions under study. When applying the wavelet decomposition, various levels of decompositions are analyzed in order to identify changes. In general, the wavelet transform of the function *f(t)* can be written in the following form [9]:

$$W\left[f(a,b)\right]=\frac{1}{|a^{1/2}|}\int_{-\infty}^{+\infty} f(t)\Psi\left(\frac{t-b}{a}\right)dt \quad (1)$$

where $\Psi\left(\frac{t-b}{a}\right)$ – the basic wavelet function; a – scaling parameter; b – shift parameter; $\frac{1}{a^{1/2}}$ – the normalizing coefficient.

During decomposition, all images are "viewed" by the selected wavelet function and joint analysis of the levels and the function "location" in the image is performed. This makes it possible to localize changes in different-time images using the correlation coefficient calculated by the formula (2) as a criterion of compliance.

$$k_{ij}=\frac{\sum_{x=i}^{N}\sum_{y=j}^{N}\left(P_i(x,y)-P_{cpi}\right)\left(P_j(x,y)-P_{cpj}\right)}{\left[\sum_{x=i}^{N}\sum_{y=j}^{N}\left(P_i(x,y)-P_{cpi}\right)^2\right]^{1/2}\left[\sum_{x=i}^{N}\sum_{y=j}^{N}\left(P_j(x,y)-P_{cpj}\right)^2\right]^{1/2}} \quad (2)$$

where k_{ij} - the correlation coefficient calculated between the levels i and j of the wavelet decomposition of the image P.

The research was conducted using high spatial resolution satellite images of Quickbird, GeoEye and WorldView. For the wavelet transform (in the discrete case), the numerical characteristics of the objects structural features in the image are matrices of coefficients with basic functions for each transformation scale level. The wavelet transformations were performed using Daubechies wavelet. Python 3.8 was used for software implementation and application of the wavelet transform for processing multispectral images.

Tree and shrub vegetation images differ from the surrounding vegetation on fallow land in brightness, tone, structure. The values of the wavelet decomposition coefficients depend on the values of the signal fluctuation, that is, in places of images where there are sharp differences in brightness values (the boundary of an object), larger values of the wavelet decomposition coefficients will be obtained than in homogeneous areas. Using this property, it is possible to localize the areas of the image, which depict land plots subject to tree and shrub vegetation overgrowth.

At the first stage of the research, the possibility of finding anomalies (tree and shrub vegetation overgrowth) on long fallow lands was determined using multispectral satellite images based on the obtained wavelet coefficients. The performed researches have shown that the Daubechies wavelet can be used as an indicator of the presence of changes in different-time images of the same area and used as a decoding feature of the image area.

In order to identify fallow lands overgrowth with tree and shrub vegetation, it is necessary to calculate the correlation coefficient between the corresponding levels of wavelet decomposition of different-time images according to formula 2. The correlation coefficient value will indicate the areas where changes have occurred.

At the second stage, researches of several images were carried out to identify changes in multi-time images. To do this, the correlation coefficient between the corresponding levels of the wavelet decomposition of images obtained on different dates was calculated. For the efficient operation of the algorithm, the image is divided into small fragments (the smallest areas) commensurate with the size of the smallest allocated overgrowth, and then a wavelet transform is performed for each smallest area (Fig. 1), this allows for further processing to use only the selected areas.

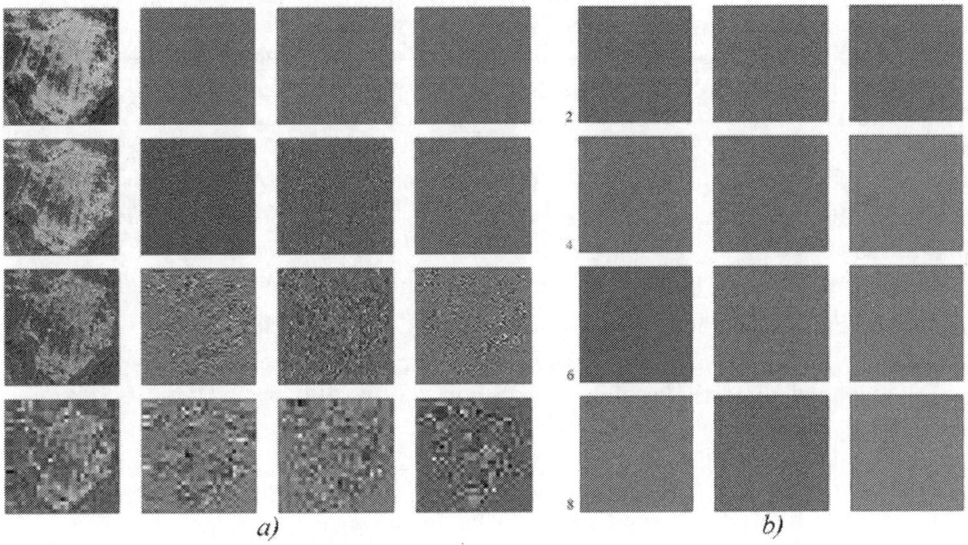

Fig1. Illustrations of image fragment Daubechies wavelet decomposition:
a) an image and its four level decomposition, b) scale transformation levels 2, 4, 6 and 8

The final stage of the technique of operational detection of unused land plots (fallow lands) subject to overgrowth by tree and shrub vegetation based on satellite images based on wavelet transformation is changes automatic vectorization in the images (Fig. 2).

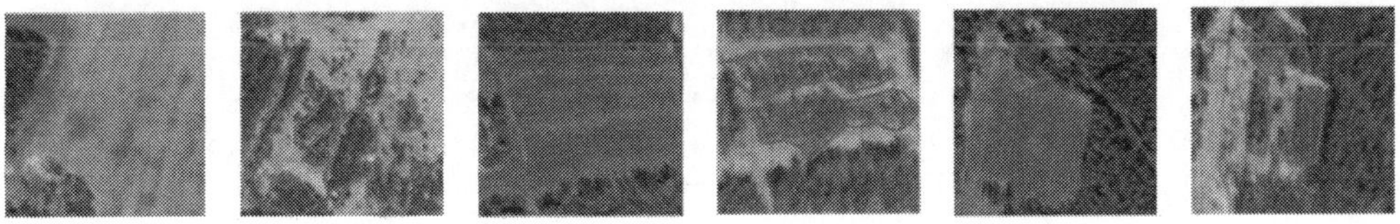

Fig. 2. Several images fragments as of 2009 and 2020 with detected overgrowth fallow by the 4th level wavelet decomposition

Then the detected changes are combined with the cartographic material at an early date in order to update outdated data.

As a result of the performed researches, it is shown that the wavelet analysis can be used at the stage of preliminary processing of multispectral satellite images in order to localize areas with fallow lands overgrowth with tree and shrub vegetation. This technique makes it possible to automate as much as possible the process of fallow lands overgrowth preliminary detection with woody and shrubby vegetation based on multi-time high spatial resolution satellite images and also to increase the processing efficiency of satellite images obtained by various surveys space systems and under different shooting conditions of the same territory.

The practical value of the proposed approach lies in the fact that the algorithm implemented in the work makes it possible to improve numerical indicators when high spatial resolution images processing of on the example of localization of fallow lands overgrowth areas with tree and shrub vegetation, and can also be used in the agricultural lands monitoring system.

References

1. D. Lu, P. Mausel, E. Brondi´Zio, E. Moran Change detection techniques, International journal of Remote Sensing, **25**, no 12, 2365-2407, (2003)
2. A. P. Guk, M. A. Altyntsev, L. G. Evstratova. Algorithms of multispectral aerospace image sequential analysis based on the use of structural-statistical approach for natural object decoding, in Proceedings of the All-Russian Conference Spatial Data Processing for Monitoring of Natural and Anthropogenic Processes, SDM 2019, 26-30 August 2019, Berdsk, Russia, 77-83, (2019). CEUR Workshop Proceedings ISSN 1613-0073, http://ceur-ws.org/Vol-2534/
3. P. Coppin, I. Jonckheere, K. Nackaerts, B. Muys Digital change detection methods in ecosystem monitoring: a review, International journal of Remote Sensing, **25**, no.9, 1565-1596, (2004)
4. C. Blatter, "Wavelets—Eine Einführung," 2nd Edition, Vieweg, Wiesbaden, 2003.
5. William K. Pratt. Digital image processing. Wilely, 4th ed, New York, (2007)
6. Gordienko A.S. The use of wavelet analysis in the processing of aerospace images. Scientific journal «Proceedings of the Higher Educational Institutions. Izvestia vuzov «Geodesy and aerophotosurveying», **2,** 68-71, (2010)
7. Marc Antonini, Michel Barlaud, Pierre Mathieu, and Ingrid Daubechies. Image coding using wavelet transform. IEEE Transactions on image processing, **1,** no. 2, 205–220, (1992)
8. Stephane G Mallat. A theory for multiresolution signal decomposition: the wavelet representation. IEEE transactions on pattern analysis and machine intelligence, **11**, no.7, 674–693, (1989)
9. Christopher Torrence and Gilbert P Compo. A practical guide to wavelet analysis. Bulletin of the American Meteorological society, **79**, no.1, 61–78, (1998)

CROP AND SOIL TEMPERATURE DIFFERENCE AN ADDITIONAL FACTOR FOR ANALYSIS OF THE CONDITION OF CROPS

Dmitriy Emelyanov[1*], *Irina* Botvich[1,2] and *Anatoly* Shevyrnogov[1]

[1]Institute of Biophysics SB RAS,660036 Krasnoyarsk, Russia
[2]Siberian Federal University, 660041 Krasnoyarsk, Russia

Abstract. The study aims to study changes in land surface temperature (LST) of soil and vegetation on agricultural land planted with barley based on unmanned LST data. Simultaneously with the LST data, the spectral characteristics (NDVI) of crops were measured using the DJI P4 Multispectral. The paper shows the variability of vegetation indices and radiation temperature during the growing season. A significant relationship was found between the dynamics of NDVI and the dynamics of radiation temperature. The features of the variability of the spatial distribution of temperatures depending on precipitation are shown. The paper gives an example of a temperature map of the studied areas in the middle of the growing season, which shows the features of the spatial distribution of temperatures.

Keywords: radiation temperature, crop, thermal map, ZENMUSE XT2, DJI P4 Multispectral, CSTD

1 Introduction

Temperature is one of the most important environmental variables affecting plant physiological processes, including transpiration, leaf water potential, and photosynthesis. Minor changes in canopy temperature, which is a function of plant leaf temperature, can be measured with a thermal imaging camera. A temperature image is formed using the long-wavelength range of radiation. Remote sensing thermal imaging cameras installed on unmanned aerial vehicles (UAV) are increasingly used in precision agriculture [1] for detecting water stress and planning irrigation [2–5], as well as for plant phenotyping [6–8]. In this work, special attention is paid to identifying the relationship between the dynamics of thermal data of agricultural lands with the dynamics of their spectrophotometric characteristics.

* Corresponding author: dima9526@gmail.com

2 Objects and Methods

The object of research is agricultural land of experimental production farm (EPF) "Minino" near the village of Minino (Emelyanovsky district) and the educational experimental farm (EEF) "Minderlinskoe" near the village of Borsk (Sukhobuzimsky district) of the Krasnoyarsk region (Fig. 1). Data obtained during the 2020 growing season.

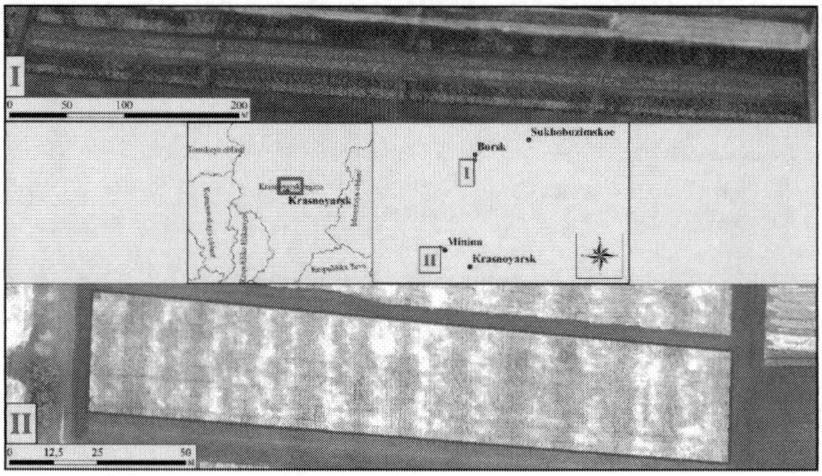

Fig. 1. Study sites: I – EEF "Minderlinskoe", II – EPF "Minino".

On the experimental field of the EEF "Minderlinskoe" grows barley variety "Acha". The fields are divided into 4 test areas in accordance with four types of treatment: deep plowing (20-22 cm) – "a"; sweep treatment (20-22 cm) – "b"; surface treatment (discator by 8-10 cm) – "c"; no tillage – "d". Fertilization was carried out along the northern part of each strip (10 m wide), the southern part remained without fertilizers.

On the experimental field of the Krasnoyarsk Research Institute of Agriculture, Federal Research Center of the KSC SB RAS, barley variety Olenek grows, sowing was carried out with deep plowing.

The work is based on data obtained by the DJI P4 Multispectral and DJI Matrice 210 RTK V2 drones with ZENMUSE XT2 thermal camera. The ZENMUSE XT2 captures images in a thermal range of 7.5-13.5μm with a resolution of 640x512 pixels. The spatial resolution of the obtained images is 12.6 cm. DJI P4 Multispectral performs multispectral imaging in 5 spectral channels (450 ± 16 nm, 560 ± 16 nm, 650 ± 16 nm, 730 ± 16 nm, 840 ± 26 nm). The spatial resolution of the obtained images is 5.3 cm. The obtained data were processed using the Pix4DMapper and QGIS software.

When thematic processing of multispectral data, the NDVI (Normalized Difference Vegetation Index) values were calculated. The CSTD (Crop and Soil Temperature Difference) was calculated as follows. For each study date, the average radiation temperature of the soil (LST_{soil}) was calculated. For this, sites without vegetation was selected as polygons in QGIS. NDVI values from DJI P4 Multispectral were used to identify areas with no vegetation. For bare soil, they range from 0 to 0.15. For visual control, an orthomosaic was used, also obtained with the DJI P4 Multispectral. Further, the values of the radiation temperature over the obtained polygons were averaged (their number varied from 30 to 50 polygons for each date). Thus, for each study date, one value was obtained, which was used in further calculations. Also, for each date of the study, the indicators of the average radiation temperature of crops (LSTcrop) were calculated separately for each type of treatment and fertilization process. These values were subtracted from the soil radiation temperature (LSTsoil - LSTcrop) to calculate the CSTD.

3 Results and discussion

The study of the relationship between the dynamics of NDVI and CSTD during the growing season was carried out. This correlation was calculated to study the relationship between these characteristics, since the dynamics of temperatures affects the vegetation process and, in turn, vegetation shielding effect affects the temperature of the soil and the retention of moisture in it.

In the initial period after sowing, the microclimate of the agrophytocenosis, in essence, did not differ from the microclimate of the black fallow, but with the emergence of seedlings, and with their further growth, significant differences arose. Due to the increasing height, foliage of plants and the crop closeness, the conditions for the penetration of radiant energy to the soil changed. The temperature regime, the intensity of turbulent exchange, the intensity of evaporation, and the humidity of air and soil changed. The vegetation cover became an active surface.

Calculation of the correlation coefficient of NDVI and CSTD average values for barley (EEF "Minderlinskoye") showed a high degree of correlation – 0.85. At the same time, for individual plots, depending on the type of treatment and fertilization, this relationship is different.

The largest correlation coefficients had data series that include measurements only with a dry soil surface. The maximum value of 0.95 is for surface treatment (fertilized ground), the minimum value is for sweep treatment (nonfertilized fround). Correlation coefficients calculated from data including measurements at dry and wet soil surface have lower values for all plots (table 1).

Table 1. Correlation coefficients of NDVI and CSTD values for barley of the EEF "Minderlinskoye", taking into account the type of processing and fertilization.

$R_{NDVI,CSTD}$	Treatment							
	fertilized ground				nonfertilized ground			
	a	b	c	d	a	b	c	d
Wet and dry soil	**0.87**	0.84	0.86	0.82	0.85	**0.77**	0.85	0.80
Dry soil	0.94	0.93	**0.95**	0.93	0.94	**0.87**	0.93	0.90

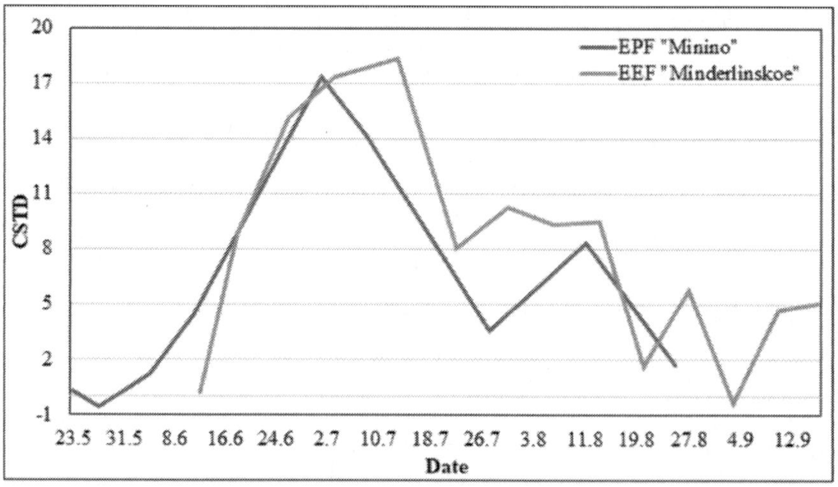

Fig. 2. Average CSTD values of barley EEF "Minderlinskoe" and EPF "Minino" during the growing season of 2020.

Comparison of the obtained data showed that the time series of CSTD crops of barley have the same character of values changes during the study period (Fig. 2).

It was found that the change in the phases of crop vegetation leads initially to an increase in CSTD, reaching maximum values on July 1 (EPF "Minino") and July 13 (EEF "Minderlinskoye"), then this value decreases. The shift in the maximum values is determined by the different sowing times of the crops. The sowing of barley at the EPF "Minino" was carried out 8 days earlier than the sowing at the EEF "Minderlinskoe".

The decrease in CSTD values on the following dates (June 12, July 22, August 20, September 3 and 10) is determined by increased soil moisture. The day before the measurements, these areas had precipitation of 4, 2, 7, 1, 2 mm, respectively. Wet soils have lower surface temperatures than dry soils. Therefore, the difference in values between soil and vegetation will be the smallest.

The general difference between the diagrams of the temperature difference at the EEF "Minderlinskoye" and the EPF "Minino" is also determined by the difference in the microclimate of the studied territories.

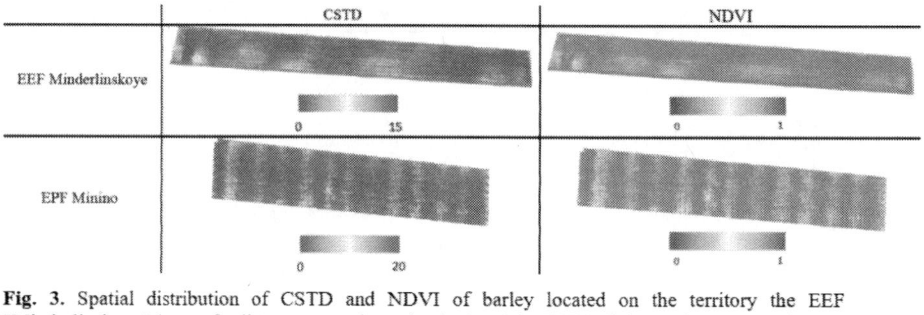

Fig. 3. Spatial distribution of CSTD and NDVI of barley located on the territory the EEF "Minderlinskoye" (type of soil treatment – deep plowing) and the EPF "Minino". The CSTD values on the territory of the EEF "Minderlinskoye" vary from 0 (min) to 15 (max) °C, EPF "Minino" from 0 to 20 °C.

The construction of thermo maps of the studied sites during the growing season made it possible to determine the features of the spatial distribution of temperatures depending on the types of soil treatment and phenological properties of crops (Fig. 3). The maps obtained show the possibility of taking into account the intra-field temperature dynamics when implementing precision farming technologies.

The study was carried out with the support of the Krasnoyarsk Regional Science Foundation within the framework of the project: "Development and implementation of a method for early forecasting of grain yield in the Siberian region according to remote sensing data of the Earth."

References

1. D.J. Mulla, Biosyst. Eng., 114 (2013)
2. C.Z. Espinoza, L.R. Khot, S. Sankaran, P.W. Jacoby, Remote Sens., **9** (2017)
3. J. Bellvert, P.J. Zarco-Tejada, J. Marsal, J. Girona, V. Gonzalez-Dugo, E. Fereres, Aust. J. Grape, R. Wine, **22** (2016)
4. V. Gonzalez-Dugo, D. Goldhamer, P.J. Zarco-Tejada, E. Fereres, Irrig. Sci., **33** (2015)
5. S. Park, D. Ryu, S. Fuentes, H. Chung, E. Hernyndez-Montes, M. O'Connell, Remote Sens, **9** (2017)
6. D. Gomez-Candon, N. Virlet, S. Labbe, A.Jolivot, J.L. Regnard, Precis Agric., **17** (2016)
7. M. Tattaris, M.P. Reynolds, S.C. Chapman, Front. Plant Sci., **7** (2016)
8. J.M. Costa, O.M. Grant, M.M. Chaves, J. Exp. Bot., **64** (2013)

CORONA imagery in the analysis of the forest-tundra ecotone in Kuznetsk Alatau (Orligtaskhyl Mountain)

Sergei Im[1,2,3*], *Il'ya* Petrov[1], and *Viacheslav* Kharuk[1,2]

[1]V.N. Sukachev Institute of Forest SB RAS (FRC KSC SB RAS), 660036 Krasnoyarsk, Russia
[2]Siberian Federal University, 660041 Krasnoyarsk, Russia
[3]M.F. Reshetnev Siberian State University of Science and Technology, 660037 Krasnoyarsk, Russia

Abstract. Changes in the forest-tundra ecotone of Orligtashyl mountain (Kuznetsk Alatau) were analyzed based on the available satellite CORONA (1971) and Wordlview (2013) imagery. Forest stands were identified on the Worldview scene using Iso cluster classification. On the CORONA scene stands were differentiated from the background area using statistical analysis of histogram of the identified classes' brightness values and were harmonized with the classification of the Wordlview imagery. Kappa-statistics showed good accuracy ($\kappa = 0.93$) of the Wordlview imagery classification and the average for CORONA ($\kappa = 0.67$). Area of stands increased by ~38% during the analyzed period. Maximal area increase was observed on the south-eastern warm slopes. Significant changes were determined on the elevational boundary of the ecotone where the stands' elevation boundary shifted up by ~9 m (0.2 m/yr.). GIS analysis revealed that trees form elongated clusters mainly directed across the slope. Observed changes occurred along with summer temperature increase during 1971–2013 by 0.3°C/decade ($r^2 = 0.32$, $p < 0.01$).

1 Introduction

An acceleration of trees' growth and trees' expansion by elevation gradient are observed since the end of the 20th century [1]. Such changes can be estimated based on the medium-resolution remote sensing data and GIS. Very high-resolution remote sensing data can improve the accuracy of assessment of the vegetation cover dynamics in the forest-tundra ecotones, these allow to account for individual trees [2]. In this paper, vegetation cover dynamics within the mountainous forest-tundra ecotone of Kuznetsk Alatau was investigated based on the CORONA and Worldview very high-resolution remote sensing data.

* Corresponding author: stim@ksc.krasn.ru

2 Materials and methods

2.1 Study area

The mountainous forest-tundra ecotone in Kuznetsk Alatau (Orligtaskhyl Mountain; 54°06′23″N; 89°04′08″E; fig. 1) was investigated. The height of the mountain is 1454 m. The administrative border between Khakassia and Kemerovskaya oblast crosses the mountain. The study area is 451 ha. In this region Siberian pine (*Pinus sibirica*) is the dominant tree species. The total amount of the annual precipitation is 560±61 mm, winter – 68±17 mm, and summer – 235±40 mm. Mean annual temperature is -3.0±1.0°C, winter – -21.0±2.6°C, and summer – +13.7±0.8°C.

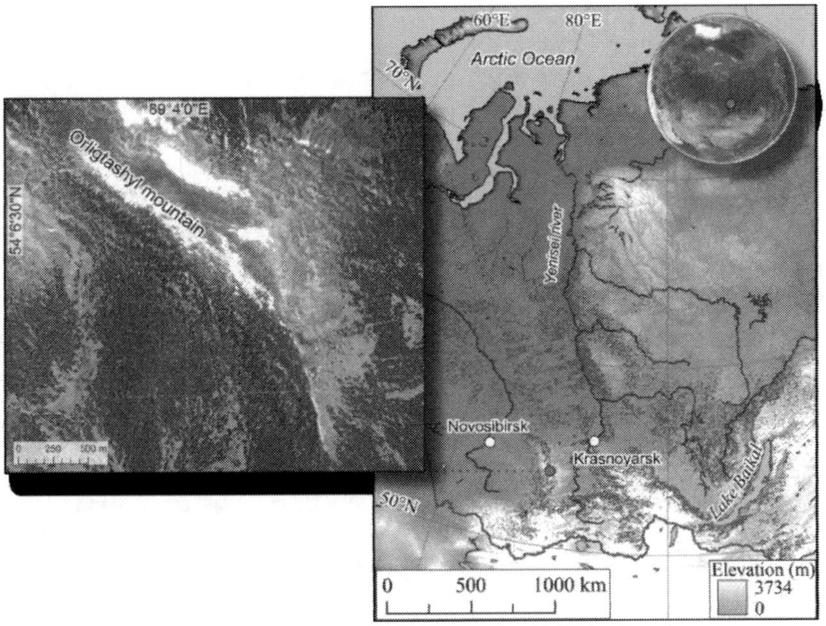

Fig. 1. Study site location (red circle) – the forest-tundra ecotone in Kuznetsk Alatau (Orligtaskhyl Mountain; 54°06′23″N; 89°04′08″E). The inserted box – Worldview-2 scene (14.07.2013) of the study area. Green on the sketch-map indicates birch stands.

2.2 Data used

2.2.1 Remote sensing data

We assessed spatio-temporal variations of the forest-tundra ecotone based on the remote sensing data received by the Corona J-3 (KH-4B satellite included to reconnaissance CORONA program (USA); scene obtained in 16.09.1971; the spatial resolution is 1.8 m) [3], and Worldview-2 (14.07.2013 г.; the spatial resolution is 0.5 m) satellites. Corona scene was acquired using USGS Earth Explorer internet geoportal (https://earthexplorer.usgs.gov). Worldview-2 scene was obtained from Bing Maps online service using SAS Planet software (http://www.sasgis.org/sasplaneta). We used digital elevation model SRTM GL1 (https://lpdaac.usgs.gov/ products/srtmgl1v003; 30 m spatial resolution) to analyze relations of vegetation dynamics with topography.

2.2.2 Ground data

Ground data were collected during the fieldwork realized in the 2016 year. In total 12 test plots were established. GPS coordinates, forest inventory, landscape, and soil data were recorded.

2.2.3 Climate data

We used climate data extracted from the CRU TS 4.05 dataset for 1971–2013 years (https://crudata.uea.ac.uk/cru/data/hrg/cru_ts_4.05). The spatial resolution is 0.5 degrees. Temperature and precipitation data were analyzed only.

2.3 Spatial analysis

A classification map based on the 2013-year Worldview scene was created using Iso cluster unsupervised algorithm [4]. Based on ground data and visual expert analysis initially created 50 classes were merged into the two final classes: (1) darkneedle stands (height > 3 m, closure > 0.2); (2) background (fig. 2). KHAT(κ)-statistics [4] was used to estimate classification accuracies based on 458 uniformly spatially distributed test points. These test points were identified based on on-ground data and expert analysis of the scene. The total classification accuracy is 95% (κ = 0.93).

The 1971-year CORONA scene was georeferenced to the 2013-year Worldview scene using splines because of considerable nonlinear geometric distortions. In total, 1107 tie points were used. Typical polygons of darkneedle stands (with height > 3 m) were selected on the grayscale CORONA scene. These polygons were used to estimate statistical variations of brightness of darkneedle stands. Computed statistical data were used to identify darkneedle stands on the CORONA scene. Pixels' brightness of darkneedle stands was in the range of 8–72. Further, the 1971-year classification map was improved by harmonization it with the 2013-year classification map. Total accuracy of the 1971-year classification map is 87% (κ = 0.67) (fig. 2).

GIS-analysis analysis was realized using ESRI ArcGIS Desktop. Microsoft Excel was used to make statistical computations.

3 Results and discussion

During the observed period (1971–2013, 42 years) area of darkneedle stands increased by 38% (0.9% per year; fig. 2). Trees' area increase was not uniform in respect to topography. Most changes were identified on the top of the mountain (1300–1400 m a.s.l.) in the forest-tundra ecotone where the area of the detected trees increased by ~150%. Elevation trees' boundary (100-highest pixels) shifted up by ~9 m (0.2 m per year). It coincides with ground measurements (0.2–0.3 m per year) [5]. This value is slightly lower than the median one for the northern hemisphere (0.35 m per year) [1]. The maximum of trees' area increase was on the southern and south-western slopes. On the north-eastern macro-slope of the mountain, trees growth ~100 m lower than ones on the south-western macro-slope. Observed changes happened along with the rise of summer (t_s) and annual (t_a) temperatures. During 1971–2013 t_s increased by 0.3°C per decade ($r^2 = 0.32$, $p < 0.01$). Mean t_s in 2000–2013 was higher on 1.0°C (U-test, $p < 0.02$) than t_s in the previous 1971–1990. During 1935–2013 t_a increased by 0.3°C per decade ($r^2 = 0.38$, $p < 0.01$). There were no significant trends of precipitation.

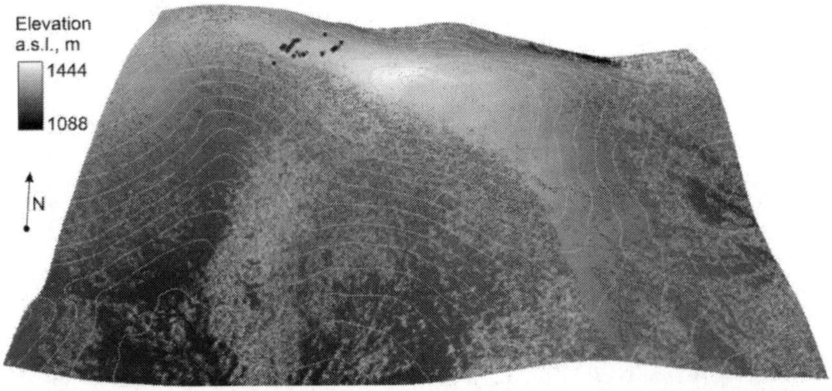

Fig. 2. The classification map of darkneedle stands identified based on the 1971-year CORONA scene (green) and the 2013-year Worldview scene (red). Blue circles indicate ground test sites. Background – SRTM GL1 digital elevation model.

A spatial statistical analysis of the average nearest neighbor analysis (centroids of which located at distance less than 190 m from each other) and multi-distance spatial cluster analysis (Ripley's K) [6] revealed that trees' clusters form groups – stripes of trees (hedges) (fig. 3a). During 1971–2013 the number of clusters increased by ~56% (1.3% per year). Clusters form stripes of trees along the north-eastern direction (~36°) which conforms to the winter winds. It is seen on the elevation boundary of the forest-tundra ecotone and proved by on-ground data (fig. 3b). Such groups of trees have resulted from the synergy impact of winds, snow, and water [1].

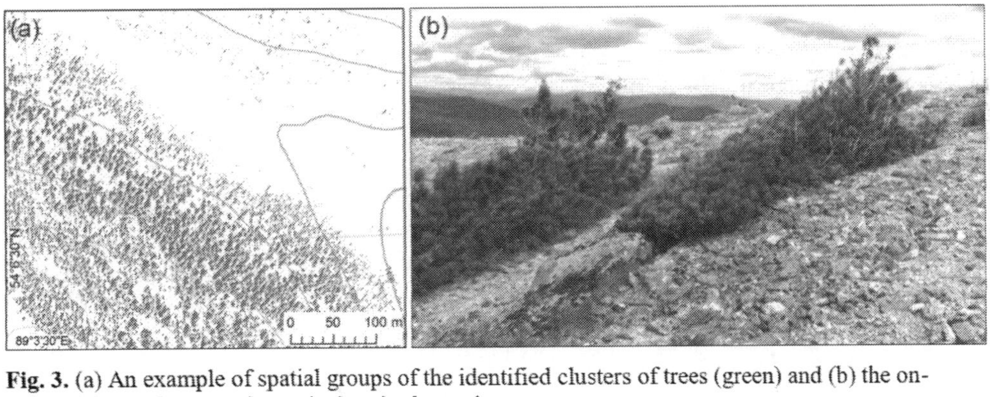

Fig. 3. (a) An example of spatial groups of the identified clusters of trees (green) and (b) the on-ground photo of *Pinus sibirica* hedges in the study area.

4 Conclusion

In this research spatio-temporal variation of forest-tundra ecotone in Kuznetsk Alatau was investigated based on the CORONA and Worldview remote sensing data. CORONA satellite data in synergy with other spatial materials allow an analysis of dynamics of the forest-tundra ecotone with sufficient accuracy. During 1971–2013 trees' area increased by 0.9% per year, and the elevation boundary of trees shifted up by 0.2 m per year. In the analyzed ecotone trees form clusters caused by the impact of winds and snow. Observed changes follow the rise of summer temperature by 0.03°C per year during 1971–2013.

The research was funded by RFBR, Krasnoyarsk Territory and Krasnoyarsk Regional Fund of Science, project numbers 20-44-240007 and 20-41-240001. Authors thank A. Golyukov and A. Shushpanov for assistance in on-ground studies.

References

1. X. Lu, E. Liang, Y. Wang, F. Babst, J.J. Camarero, Glob. Ecol. Biogeogr., **30**, 305-315 (2021)
2. V.I. Kharuk, S.T. Im, M.L. Dvinskaya, K.J. Ranson, I.A. Petrov, J. Mt. Sci., **14**, 442-452 (2017)
3. Declassified intelligence satellite photographs. Fact sheet 090-96, https://web.archive.org/web/20070705233001/http://erg.usgs.gov/isb/pubs/factsheets/fs09096.html (1998)
4. J.A. Richards, J. Xiuping. Remote Sensing Digital Image Analysis: An Introduction (2006)
5. I.A. Petrov, A.S. Shushpanov, A.S. Golyukov, V.I. Kharuk, Sib. J. For. Sci., **5**, 43-53 (2019)
6. P.M. Dixon, *Ripley's K Function*, in Wiley StatsRef: Statistics Reference Online (2014)

The methods and automatic technology aimed at imagery georeferencing, cloud screening, atmospheric and radiometric correction of KMSS-M satellite data

Pavel Kolbudaev[1], *Dmitry* Plotnikov[1*], *Evgeny* Loupian[1], *Andrey* Proshin[1] and *Alexey* Matveev[1]

[1]Space Research Institute of the Russian Academy of Sciences, 117997 Moscow, Russia

Abstract. In this study we present methods and automatic technology developed for routine processing of satellite imagery acquired by cameras MSU-201 and MSU-202 (KMSS-M) onboard Meteor-M №2. The developed methods were aimed at imagery georeferencing issues fixing, clouds and shadows detection as well as atmospheric and radiometric correction. Basing on these methods we built an automatic technology and complete KMSS-M data processing chain which provided analysis ready dataset for Russian grain belt and adjacent areas of neighboring countries for the year 2020. Method for imagery georeferencing was based on Pearson's correlation localized maximization when compared to the georefenced and cloudfree coarse-resolution reference image produced in IKI RAS through MOD09 product time series processing. Method for clouds and shadows detection was based both on the spatial analysis of outputs from geocorrection step and auxiliary image, characterizing georeferenced KMSS-M image values relative accordance with the IKI reference image. The atmospheric correction was based on localized histogram matching of KMSS-M and IKI reference date-corresponding imagery, and thereby concurrently performed radiometric correction of KMSS-M data, compensating effects of varying viewing and illumination geometry which explicitly manifest across 960-km-wide swath area. The developed methods are noticeably minimalistic, requiring only one target spectral band to perform properly. Due to high flexibility and robustness, they also may be applied to raw satellite imagery acquired from various Earth observation systems, including Russian systems of high and moderate spatial resolution. The technology is currently being deployed in an operative mode for several test sites of Russia since the year 2021 onwards.

Meteor-M No. 2 meteorological satellite, which carries KMSS-M instrument (acronym for multispectral satellite imagery complex), was launched on July 8, 2014 and placed on sun-synchronous orbit at an altitude of about 825 km [1]. The KMSS-M consists of two cameras, designated MSU-201, MSU-202, and one additional camera, designated MSU-

* Corresponding author: dmitplot@d902.iki.rssi.ru

250. In this study we focus on data acquired by two multispectral cameras MSU-201(2). The cameras are optimized for land monitoring and operate in three spectral bands – green (0.535-0.575 microns), red (0.630-0.680 microns) and near infrared (0.760-0.900 microns) with a spatial resolution of about 60 meters, the formed viewing band from two cameras is about 960 km [1]. The CCD elements of the corresponding spectral channels are separated in the focal plane in the direction of satellite's orbital motion, which leads to interband parallax, evident on elevated objects such as clouds, mountain peaks, as well as to a time lag between recording one objects' reflectance in different spectral bands. KMSS-M imagery is characterized by a rare combination of spatial resolution and acquisition repeatability, so Russia is completely covered with KMSS-M imagery every 2-3 days. The launch of the Meteor-M satellite No. 2-2 on July 5, 2019 with the following KMSS device onboard provides full coverage of Russia with virtually daily acquisitions.

The developed methods of georeferencing, cloud and shadow screening, as well as radiometric and atmospheric correction are based on spatial analysis and localized comparison with a date-corresponding coarse-resolution reference image, produced in IKI RAS using reconstruction methods [2, 3]. The reference image contains measurements of surface reflectance reconstructed data for required date and is free from hindering factors (clouds, shadows, snow cover). MODIS reference images were obtained by interpolating a time series of MOD09 product using LOESS weighted interpolation approach [2], where normalized values of the NDSI index, sensitive to the influence of interfering factors, were used as weight factors. With Sentinel-2 (MSI) high resolution imagery archive for entire territory of Russia for the year 2020, it was found that reconstructed reference is better geolocated compared to the original MOD09 product, featuring average geolocation error of about 15 meters (about 6% of the pixel size).

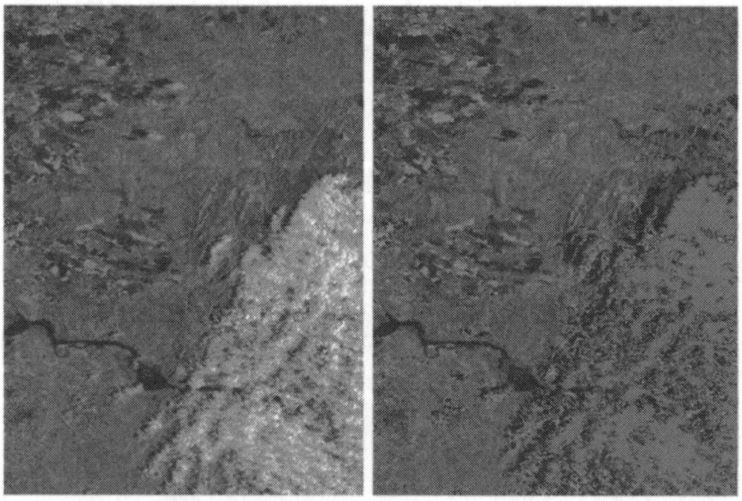

Fig. 1. Part of clouded MSU-201 scene (NIR band) for May 15, 2020. Detected clouds are marked red, shadows are marked blue

Two-staged approach for georeferencing of KMSS-M data is based on an optimized search of possible positions of the MSU-201(2) image relative to the reference image and finding an spatial offset at which the value of Pearson's correlation between data is maximized. At the first stage the systematic geolocation bias of the MSU-201(2) image is eliminated through Pearson's correlation maximization using the whole scene data. At the second stage geocorrection process operates with small local areas (blocks) of the original image. When calculating correlation, the portions of the KMSS-M images are sequentially coarsed in a sliding window of 4 by 4 pixels and the result of coarsening is compared with the corresponding pixel of MODIS reference low-resolution image. The search for the optimal position of the block is performed with a spatial stepping equal to one KMSS-M pixel. The distinctive feature of this method is the ability to geocorrect cloudless areas of high spatial resolution images using low spatial resolution reference in separate spectral bands independently. The comparison of geo-corrected KMSS-M images with the collection of cloudless Sentinel-2 (MSI) images for entire territory of Russia for the 2020 year indicates that average geolocation error does not exceed 40 meters.

The cloud detection method is based on the use of the maximum values of Pearson's correlation, stored as an image, obtained at the stage of georeferencing and which characterizes the presence and degree of surface hindering factors. Such image is threshold-classified into two classes - mostly clear and mostly clouded areas. Comparison of the KMSS-M image with MODIS reference and the use of the described auxiliary binary classification enable to set a local adaptive local threshold for cloud detection and thus calculate cloud masks.

The shadow masking method is based on Sun illumination geometry reconstructed using geographic coordinates and image acquisition timestamps. The area of potential cloud shadows is determined using calculated scene illumination geometry and auxiliary image of the difference between KMSS-M and the reference image. Figure 1 shows an example of highlighting clouds and shadows on a part of the MSU-201 for May 15, 2020. The advantage of the method is the ability to detect clouds and shadows using raw and uncalibrated data in separate spectral bands independently.

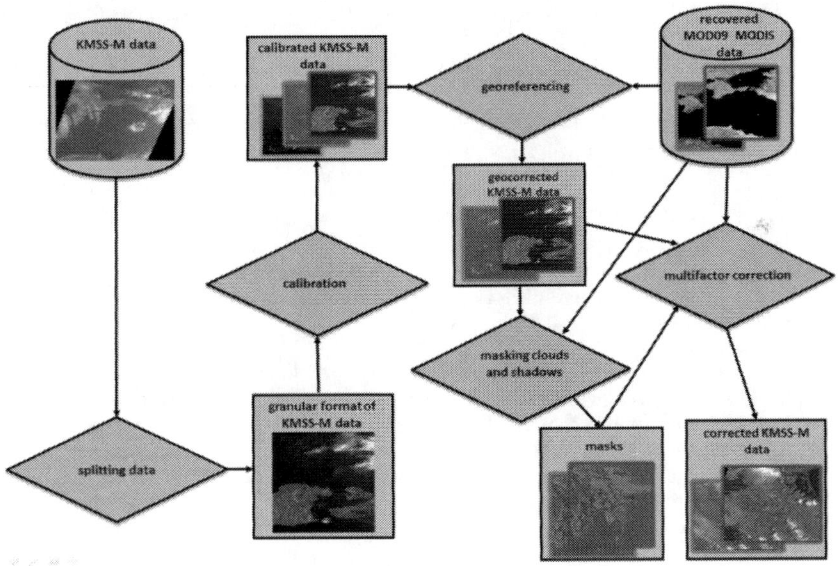

Fig. 2. General flowchart of KMSS-M data processing technology

The method for radiometric and atmospheric correction of KMSS-M images is based on localized and joint spatial analysis of cloudless image areas with the MODIS data. In the process of joint spatial analysis a variable-size sliding window collects local statistics of the required volume characterizing the localized distribution of brightness values in the compared data sources. Because of the area of analysis, date of the reference and the KMSS-M fit the resulting distributions both histograms have the same form and can be match to each other on the assumption that the influence of the atmosphere shifts and deforms the histograms of the brightness distribution [4]. For each local sliding window centered in the local area of the image (block) linear transformation coefficients are automatically calculated and applied to the pixels of window. As a result each pixel of the image being corrected and receives many estimates from its own and surrounding blocks with level of accuracy of estimates. The combination of all estimates makes it possible to convert the initial values on the KMSS-M to the values of the spectral brightness of the reference image.

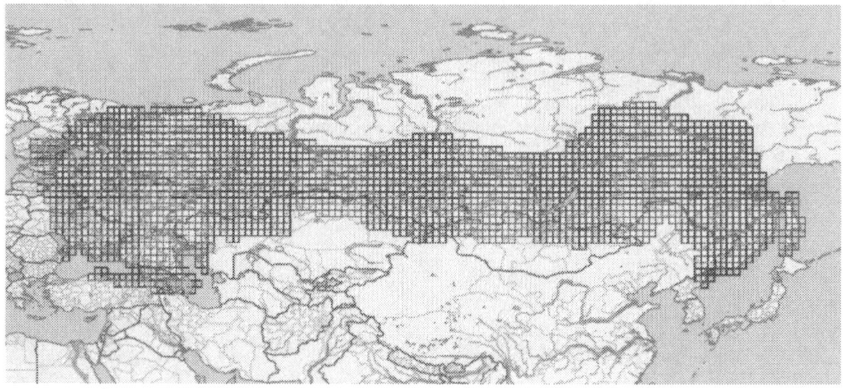

Fig. 3. Availability map for KMSS-M standard products acquired during the year 2020 with Eurasian continent on background

The method of radiometric calibration and atmospheric correction of KMSS-M images suppresses radiometric effects which appear due to a significant difference in observation and illumination angles across 960 km of swath. The advantage of the described method of multifactor correction is the ability to process each spectral band independently.

The developed methods of georeferencing, clouds and shadows detection, as well as radiometric and atmospheric correction were used as the basis for the automatic technology for routine data processing of KMSS-M imagery deployed at the IKI RAS [5]. The technology implements spatial re-arrangement of the data from scenes to a granular form together with more convenient naming for all derived products that are more convenient for their spatial-temporal analysis, cataloging and distribution. In the approved standard the file names include information about the product name, date (year, day and time), sensor ID and granule number. The flowchart of the developed technology is shown in Fig. 2. The results of the technology are available in the cartographic interface of the Sozvezdie-Vega family systems (sozvezdie-vega.ru) and include a standard set of products: atmospherically corrected and georeferenced KMSS-M images (Red and NIR bands), cloud and shadow masks for each spectral channel separately as well as NDVI [8]. These products are processed for the areas of Russian grain belt and neighboring countries and available for download and analysis. The data processing is carried out using the resources of the collective use center «IKI-Monitoring" [6].

References

1. B. Zhukov, T. Kondratieva, I. Polyanskiy, *Interannual sensitivity trend of the cameras of the multispectral satellite imaging system KMSS-M on Meteor-M No. 2 spacecraft based on the in-flight calibration in 2015–2020*, Sovremennye problemy distantsionnogo zondirovaniya Zemli iz kosmosa, Vol. **18**, No. 1, pp. 53-60 (2021)
2. W. Cleveland, *Robust Locally Weighted Regression and Smoothing Scatterplots,* American Statistical Association, V. **74**, 368, P. 829–836 (1979). DOI: 10.2307/2286407.
3. D. Plotnikov, E. Elkina, E. Dunaeva, S. Khvostikov, E. Loupian, S. Bartalev, *Development of the Method for Automatic Winter Crops Mapping by Means of Remote sensing Aimed at Crops State Assessment over the Republic of Cremia*, Taurida Herald of the Agrarian Sciences, No. 1, **21,** P. 64-83 (2020). DOI 10.33952/2542-0720-2020-1-21-64-83. 7.
4. D. Hadjimitsis, G. Papadavid, A. Agapiou, K. Themistocleous, M. Hadjimitsis, A. Retalis, S. Michaelides, N. Chrysoulakis, L. Toulios, C. Clayton, *Atmospheric correction for satellite remotely sensed data intended for agricultural applications: impact on vegetation indices*, Natural Hazards and Earth System Sciences, V. **10,** P. 89–95 (2010)
5. D. Plotnikov, P. Kolbudaev, B. Zhukov, A. Matveev, S. Bartalev, V. Egorov, A. Kashnitskiy, A. Proshin, *The collection of multispectral KMSS-M (Meteor-M No. 2) satellite data aimed at quantitative assessment of the Earth surface*, Sovremennye problemy distantsionnogo zondirovaniya Zemli iz kosmosa, Vol. **17,** No. 7, P. 256–262 (2020). DOI: 10.21046/2070-7401-2020-17-7-256–262
6. E. Loupian, A. Proshin, M. Bourtsev, A. Kashnitskiy., I. Balashov, S. Bartalev, A, Konstantinova, D. Kobets, A. Mazurov, V. Marchenkov, A. Matveev, M. Radchenko, I. Sychugov, V. Tolpin, I. Uvarov, *Experience of development and operation of the IKI-Monitoring center for collective use of systems for archiving, processing and analyzing satellite data*, Sovremennye problemy distantsionnogo zondirovaniya Zemli iz kosmosa, Vol. **16,** No. 3, P. 151-170 (2019). DOI: 10.21046/2070-7401-2019-16-3-151-170

Arctic lakes thickness measurement in the Lena River delta based on the Sentinel-1 observations

Yuliya Leskova[1], *Konstantin* Muzalevskiy[2*], and *Zdenek* Ruzicka[2]

[1]Siberian Federal University, , Krasnoyarsk, Russian Federation
[2]Kirensky Institute of Physics of the Siberian Branch of the RAS – Division of Federal Research Center "Krasnoyarsk Scientific Center of the Siberian Branch of the RAS", Krasnoyarsk, Russian Federation

Abstract. In this paper, we investigate the possibilities of ice thickness remote sensing on freshwater lakes based on Sentinel-1 observations. Fr. was selected as a test site. Kurungnakh Island, located in the Lena River delta, was selected as a test site, where 10 lakes with floating and grounded ice were selected. For the test lakes, the time series of backscatting coefficients measured by Sentinel-1 at horizontal-horizontal polarization (EW mode) and at an angle of 18-20 from September 2015 to June 2017 were analyzed. The average lakes ice thickness was estimated based on the Lebedev's model and air temperature according to the European Center for Medium-Range Weather Forecasts (ECMWF ERA5). It is shown that, the dependence between backscattering coefficient and ice thickness (120-140 cm) can be described a linear function (with a standard deviation of 13.5-25.3 cm and a determination coefficient of 0.70-0.79). The preliminary studies shows that the Sentinel-1 radar data is promising for remote sensing of ice thickness on freshwater arctic lakes.

Introduction

Currently, the Sentinel-1 radar data is not used for remote sensing of the ice thickness on Arctic freshwater lakes. At the same time, the the ice lake thickness is a sensitive indicator, reflected the state of the climate system, and an esential climatic variable, recommended by the World Meteorological Organization for global observation [1]. This paper investigates the possibilities of ice thickness remote sensing for a group of freshwater lakes located on the Kurungnakh Island (Lena River Delta) based on Sentinel-1 radar data (operats on a frequency of 5.4 GHz, wavelength is about 5.6 cm). The backscatter coefficient (RBC) over test lakes was measured by the Sentinel-1 satellite at HH-polarization (horizontal-horizontal) at an angle between 18-20 degrees from September 2015 to June 2017 in EW mode. Radar images were processed in Sentinel Application Platform (SNAP) according to the standard procedure: applying precise orbits, radiometric calibrate, thermal noise removal, speckle filtering (5x5 Lee filter), range Doppler terrain correction.

* E-mail: rsdkm@ksc.krasn.ru

1 Modeling lake ice growth

Due to the fact that the ice thickness h was not measured at the radar observations, the empirical Lebedev's model [2] was used, expressed by the following equation

$$h=1,33(\Sigma T_a)^{0,58} \text{ [cm]}, \quad (1)$$

here $\Sigma|T_a|$ sum of modules of daily negative air temperatures (in degrees Celsius). To calculate T_a, the air temperature (on a height of 2 m) was used from the European Center for Medium-Range Weather Forecasts (ECMWF ERA5, spatial resolution 0.25°) [3]. As an example, Fig. 1 shows the calculations of the lake ice thickness in the area of the southern part of the Kurungnakh Island (72.2987 N, 126.1785 E) based on the Lebedev's model with according to ECMWF data for the period 2016-2017. Note that a strong correlation (determination coefficient, R^2, is about 0.979) and a small error (root-mean square error, RMSE = 2.15°C) between the air temperature calculated on the basis of ECMWF data and measured by the Stolb Island weather station [4], located several kilometers south east of Kurungnah Island (see Fig. 2), which allows to use of ECMWF data to estimate air temperatures in the Arctic regions, where weather station data are not available.

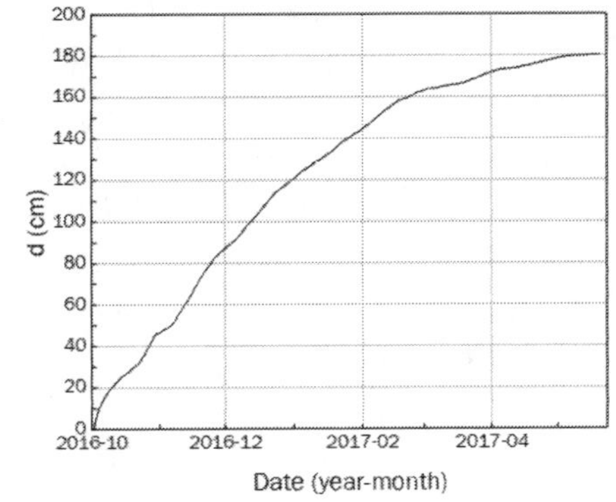

Fig. 1. Ice thickness, calculated based on Lebedev's model, for southern part of Kurungnakh Island (2016-2017).

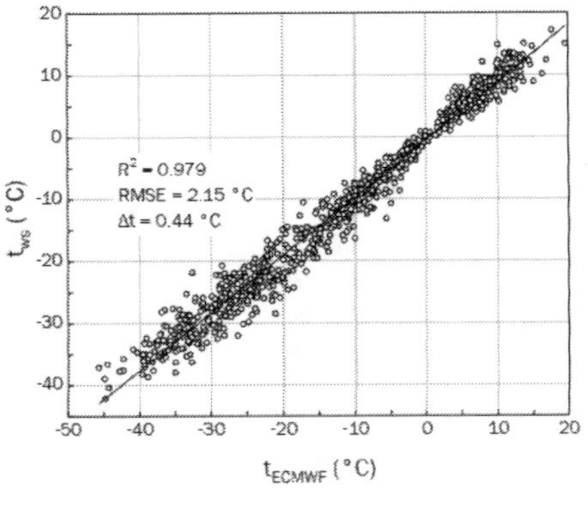

Fig. 2. Daily air temperature between weather station data Stolb Island (T_{ws}) and ECMWF ERA5 (T_{ECMWF}).

2 Method for assessing ice thickness

The dependence betwen BC measured by Sentinel-1 for a group of shallow lakes (1.05–1.8 m deep [5, see Table 2]) and the ice thickness calculated on the basis of the Lebedev's model (see Fig. 1) is shown in Fig. 3. It can be seen (Fig. 3) that the decrease in BC (occurring due to the contact of the ice to bottom of the lakes) is in good agreement with the depth d_i of the lakes [5, see Table. 2]. From the data shown in Fig. 4, it follows that the ice thickness of the lakes freezing to the bottom, determined on the basis of Sentinel-1 data and Lebedev's model (at the moment when ice touches the lake bottom) with RMSE = 7.4 cm and R^2 = 0.934 agrees with ground-based measurements [5, see table. 2]. As a result, it was shown that Lebedev's model can be used to predict the ice lake thickness with an error of no better than 7.4 cm.

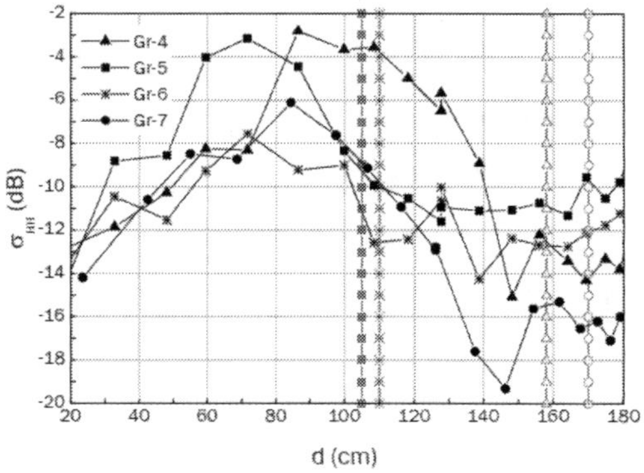

Fig. 3. BC vs ice thickness. The vertical dashed lines mark the depths of the corresponding lakes (highlighted by a symbol) according to the data [5, see table. 2].

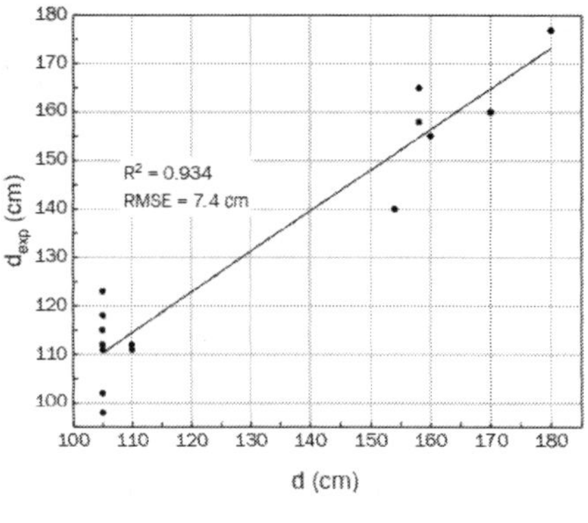

Fig. 4. Dependence between the ice thickness d_{exp} of lakes freezing to the bottom, estimated according to Sentinel-1 data and Lebedev's model, and the depth of the lakes according to [5, see Table. 2].

In the case of lakes that do not freeze to the bottom, there is a linear relationship between the ice thickness (estimated based on the Lebedev model) and BC, measured by Sentinel-1 at HH polarization (see Fig. 5). During two seasons 2015-2016 and 2016-2017, the RMSE in predicting the lake ice thickness (see [5, Table 2]) based on the BC measured by Sentinel-1 varied from 13.5 to 25.3 cm. The R^2 varied from 0.700 to 0.785 (Fig. 5).

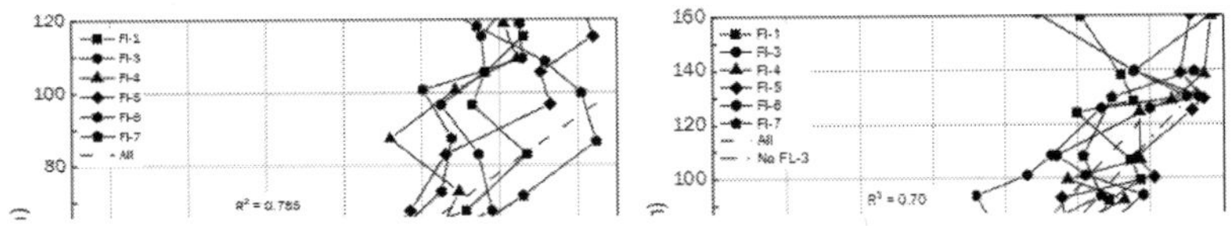

Fig. 5. Dependence between ice thickness calculated on the basis of the Lebedev's model and BC measured by Sentinel-1 at HH polarization in winter: a) 2015-2016 and b) 2016-2017.

Conclusion

Sentintel-1 data allows to identifying floating and grounded ince in Arctic lakes in the Lena River delta (Kurungnah Island). As a result, on the basis of Sentinel-1 radar data, a remote technology can be proposed for assessing the fish resources of the Arctic lakes, the availability of unfrozen water for industrial needs, and monitoring of climatic changes in the Arctic zone. It is shown that lake ice, up to 120-140 cm thick, can be determined based on Sentinel-1 data with an error in the range 13.5-25.3 cm.

References

1. *The Global Observing System for Climate: Implementation Needs, GCOS-200,* GCOS 2016 Implementation Plan, World Meteorological Organization, 315, Geneva, Switzerland (2016)
2. Lebedev V.V. *Ice growth in Antarctic rivers and seas depending on negative air temperatures.* Problemy Arktiki –Atctics Problems, **5-6**, 9-25, (1938) (In Russian)
3. European Centre for Medium-Range Weather Forecasts. Режим доступа: https://www.ecmwf.int/
4. Stolb weather station data. Available online: https://rp5.lv/Архив_погоды_на_о._Столб
5. S. Antonova, C. R. Duguay, A. Kaab, B. Heim, M. Langer, S. Westermann, J. Boike, *Monitoring Bedfast Ice and Ice Phenology in Lakes of the Lens River Delta Using TerraSAR-X Backscatter and Coherence Time Series*, Remote Sens, **8(903)**, 6, (2016)

Forests burnability in Siberian subregion and post-fire effects according to modern and retrospective satellite surveys

Anna Malkanova [1,2]*, *Andrey* Zabrodin [1,2], *Evgenii* Ponomarev [2,3]

1 Federal Research Centre "KSC SB RAS", 660036, st. Akademgorodok, 50, Krasnoyarsk, Russia
2 Siberian Federal University, 79/10 Svobodny Ave., 660041, Krasnoyarsk, Russia
3 V.N. Sukachev Institute of Forest SB RAS, 660036, st. Akademgorodok, 50/28, Krasnoyarsk, Russia

Abstract. The dynamics of fires in the territory of Eastern Siberia (the territory of the Lena River basin in the middle reaches) has been studied with the use of materials from retrospective satellite imagery. The dependence of the burnability of forests in the local territories of Eastern Siberia on the level of heat and moisture supply is shown. The range of the burnability index (the number and area of fires) for the study area under extreme drought conditions was revealed, confirmed by retrospective data. It is shown that potentially the average annual values of the number of fires can be 2.5 times higher than the current statistics. Based on the invariant NDVI and NDWI indices, a range of changes in the characteristics of post-fire areas has been identified, which indirectly determines the level of fire impact on vegetation and the humidification regime.

Introduction

The problem of forest fires is relevant all over the world. Over the past 25 years, satellite monitoring has made it possible to assess the scale of fire processes in the forests of Siberia [1-3]. This is important for forecasting fire regimes in certain subregions in changing climatic conditions [4-6]. In order to increase the reliability of such forecasts, the series of available instrumental data on wildfires should be expanded using available retrospective satellite imagery (Fig. 1) [7].

Episodic data on fires in Siberia can be found in the materials of Meteor satellites obtained in the 1980s (Fig. 1,1), and modern data from Terra/MODIS satellites can also be found (Fig.1.2). Based on a preliminary analysis of these materials, we formulated a hypothesis that forecasting the dynamics of the burnability and probable scenarios of fire regimes in the near future requires prolonged information about fires from sources such as reliable instrumental satellite observations since the early 1970s.

The survey material for the 70–80th years of the last century on separate subregions of Siberia partially available according to the Landsat 4–5 data (https://earthexplorer.usgs.gov/) (Fig. 1.3).

The purpose of this research paper is to analyse the connection between the burnability and meteorological parameters on the example of Eastern Siberia (the middle part of the Lena River basin) based on the characteristics of the extreme season in 1986, as well as assessment of the level of wildfire effects on the vegetation cover and on the dynamics of the spectral characteristics of disturbed areas.

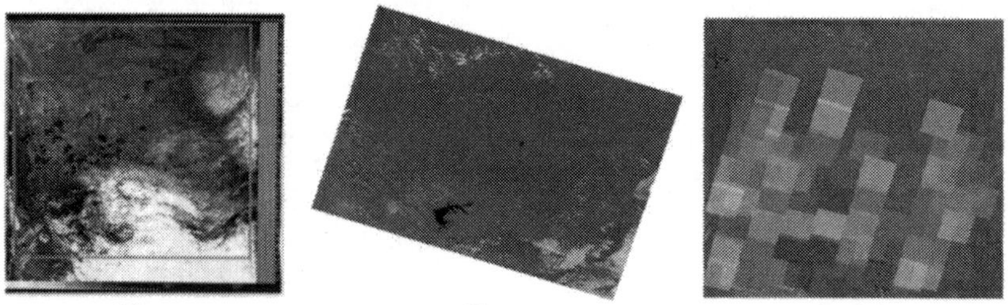

1) 2) 3)

Figure 1. Research area - the Lena River basin area in the middle reaches 1) Satellite image Meteor, 1986, 2) Satellite image Terra/MODIS, 2020, 3) Available satellite scenes Landsat 4-5 TM, 1986. The red square marks the border of the research area

Source data and processing methods

The research paper analysed data on post-fire areas recorded in Eastern Siberia (Lena River basin, 58–61°s.w., 118–122°v.d.) for the period 1986–2015 based on the materials of USGS [8] satellites Landsat 1–8 MSS/TM (Multispectral Scanner Sensor/Thematic Mapper) (Fig.2). Taking into account all available personnel, the percentage of coverage of the area of interest is not less than 65%.

We studied groups of images obtained from Landsat 4–8 MSS C1 satellites for four periods: 1986, 1995, 2008, 2015 in a combination of red (0.61–0.76 μm), green (0.54–0.57 μm) and blue (0.48–0.51 μm) channels. The spatial resolution is 30 m. The sample consisted of 12 images (3 images for each year). The total area of the research ground is 81.53 thousand km^2.

With the help of the QGIS software package (Quantum Geographic Information System, version 3.20), a series of images was classified with training (the module in QGIS dzetsaka: Classification tool, https://github.com/nkarasiak/dzetsaka/).

After the classification, the areas of fire ranges and the indicator of burnability (Relative Burnt Area) were calculated as the ratio of the total area of fires per year to the forest area of the territory [8]:

$$RBA = \frac{\sum_{i=1}^{n} S_{P_i}}{S_{gen.}},$$

where S_{P_i} is the area of a single fire, $S_{gen.}$ is the total area of the researched ground.

To assess the dependence of the burnability of the forests of the research area on the level of heat and moisture availability of the territory, the Selyaninov hydrothermal humidification coefficient (HTC) was calculated [9]:

$$\text{HTC} = 10 \cdot \sum_{i=1}^{n} \frac{R_i}{t_i},$$

where R_i is precipitation in millimeters at t >+10°C for the i-th day of the season, мм, t_i is the temperature value in the same period, °C, 10 is the normalization coefficient.

To assess the characteristics of post-fire areas, the Normalized Difference Water Index (NDWI) was used, which indirectly characterises the moisture content in plants and soil [10]:

$$NDWI = \frac{NIR - SWIR1}{NIR + SWIR1}$$

where NIR (λ = 0.63–0.69 μm) and SWIR1 (λ = 1.55–1.75 μm) are the reflection coefficients measured in the specified wavelength ranges.

Additionally, the Normalized difference vegetation index (NDVI) was calculated to assess the state of vegetative vegetation covers for each season of the study period. In accordance with generally accepted criteria, we used the values of this indicator to assess the state of vegetation after fires, as well as for the general characteristics of the disturbance of the research area in certain seasons. The index was calculated based on the ratio:

$$NDVI = \frac{NIR - RED}{NIR + RED},$$

where NIR (λ=0.63–0.69 μm) and RED (λ=0.76–0.90 μm) are the reflection coefficients measured in the specified wavelength ranges.

Figure 2 shows an example of the distribution of vegetation indices NDVI (Figure 2.2) and NDWI (Figure 2.3) for 1986. To calculate this index, a standard methodology was used based on channel #3 and channel #4 for Landsat 1–7 or channel #4 and channel #5 for Landsat 8 [11,12].

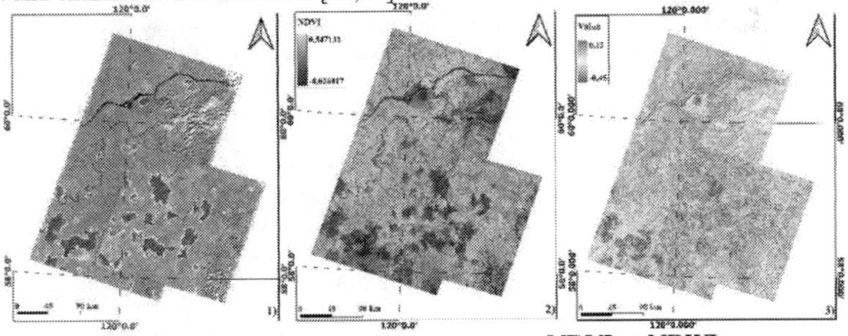

Figure 2. Landsat 1–5 MSK 1 survey for 1986: 1) Selected post-fire areas, 2) NDVI, 3) NDWI

Results and discussion

The level of extreme burnability of the studied subregion in the 1986 fire season was recorded (RBA=6.96%). This indicator is significantly (10 times or more) higher than the average long-term level of burnability over the past 25 years, which was determined based on the materials of instrumental satellite monitoring of fires conducted in Siberia since 1995 (Table) [4].

Table. The results of processing satellite monitoring materials for the research area for 1986, 1995, 2008 and 2015.

Year	*Number of fires*	*Area of fires, km^2*	*RBA, %*	*HTC*	*$NDVI_{medium}$ (Fires/ Not damaged)*	*$NDWI_{medium}$ (Fires/ Not damaged)*
1986	66	5680.96	6.96	0.36	0.03/0.5	–0.51/0.07
1995	20	156.76	0.19	1.09	0.3/0.6	–0.33/0.15
2008	40	706.86	0.87	1.15	0.15/0.65	–0.39/0.18

Year	*Number of fires*	*Area of fires, km²*	*RBA, %*	*HTC*	$NDVI_{medium}$ *(Fires/ Not damaged)*	$NDWI_{medium}$ *(Fires/ Not damaged)*
2015	17	134.82	0.17	0.74	0.02/0.2	–0.09/0.2
Average without 1986*	26±13	332.81± 366.77	0.41± 0.45	0.99± 0.25	0.16±0.14/ 0.48±0.25	–0.27±0.16/ 0.8±0.03

* Data on the extreme burnability of the 1986 fire season are not included, as they have a significant excess relative to the average annual norm.

Prolongation of the fire chronology for 10 years made it possible to identify the burnability of the area of interest in the conditions of the extreme fire season of 1986. Compared with the subsequent period in such a season, the number of fires is 2.5 times higher, and the total area of fire damage (in terms of RBA) was up to 7%. This allows us to make an assumption about the possibility of implementing extreme fire seasons (with a level of burnability (RBA_{EXTR} = 6.96%) significantly higher than the average long-term (RBA=0.41±0.45)) in the region. This situation is likely to happen in the future, taking into account the changing climate and the increasing aridity of the growing season (according to indirect data of the HTC [4]).

An indirect assessment of the vegetation state of disturbed territories was obtained by analyzing invariant NDVI and NDWI indicators in comparison for post-fire sites and background (undisturbed) territories. It was found that after the fire, the value of the NDVI indicator decreases by up to 75% relative to the norm (undisturbed adjacent areas). This is consistent with the results of studies for other objects [13, 14]. An additional characteristic was obtained based on the results of processing NDWI data, which characterizes the degree of moisture availability of sites. The results showed that after fires, the NDWI value changes upward, and relative anomalies compared to the background are 350%.

Conclusion

We discussed the prospects of using archived satellite monitoring data to analyze the burnability and post-fire changes.

We found that the burnability correlates (r = –0.83) with the Selyaninov hydrothermal humidification coefficient (HTC), which decreased to 0.36 under the conditions of extreme drought in 1986 with an average long-term optimum for the region of ~0.99±0.25.

Significant deviations in the positive direction have the NDVI and NDWI indices. The standard values of these indices may affect the forecast of plant moisture supply. With values NDVI = 0.03 ... 0.71 and NDWI = 0.07, we have a high HTC index equal to 0.74, which corresponds to low burnability (RBA = 0.17) in 2015.

It is shown that in conditions of increasing the level of aridity of the growing season, a repetition of the scenario of the extreme fire season of 1986 is possible. At the same time, the average long-term indicators of burnability (RBA = 0.41±0.45) can be significantly exceeded, as was observed in the example of 1986, when the RBA reached 6.96%.

The work was carried out on the topics of state assignment No. 0287-2019-0006 and No. 0287-2021-0040.

Acknowledgements. This work was performed using the subject of project of IF SB RAS no. 0287-2021-0008 and FRC KSC SB RAS no. 0287-2021-0040. This research was partly funded by the Russian Foundation for Basic Research (RFBR) and Government of the Krasnoyarsk krai, and Krasnoyarsk krai Foundation for Research and Development Support, no. 20-44-242002 ("Instrumental monitoring of physical properties and numerical modeling of the state of technogenically disturbed soils in Siberia").

References

1. V.I. Kharuk, E.I. Ponomarev, G.I. Ivanova, M.L. Dvinskaya, S.C.P. Coogan, M.D. Flannigan, Ambio. 50. pp.1953–1974. (2021) https://doi.org/10.1007/s13280-020-01490-x
2. S.A. Bartalev, F.V. Stytsenko, Lesovedenie. № 2. pp. 115–122. (2021)
3. A.I. Sukhinin, Sibir. jekologich. zhurn. **3**. pp. 85–91. (1996)
4. E.I. Ponomarev, A.S. Skorobogatova, T.V. Ponomareva, Meteorology and Hydrology, **43**, pp. 456–463. (2018) DOI: 10.3103/S1068373918070051.
5. S.G. Conard and E.I. Ponomarev, Wildfire. **29**. pp. 26–32. (2020) ISSN: 1073-5658.
6. W. Min Hao, B.L. Nordgren, A. Petkov, R.E. Corley, S.P. Urbanski, M.C. Reeves, L. Scott Baggett, Y. Balkanski, P. Ciais, F. Mouillot, C. Yue, Biogeosciences. **18**. pp. 2559-2572. (2021)
7. A.V. Malkanova, A.N. Zabrodin, N.D. Yakimov, E.I. Ponomarev, Mat. III Vseros. nauch.-prakt. konf. «Aktualnye problemy obespechenija pozharnoj bezopasnosti i zashhity ot chrezvychajnyh situacij». Zheleznogorsk, 23 April 2021 g., Zheleznogorsk: SPSA GPS MChS Rossii, pp. 346–351. (2021)
8. V.V. Usenja, E.N. Katkova, S.V. Uldinovi, *Lesnaja pirologija: uchebnoe posobie dlja studentov vysshih uchebnyh zavedenij po special'nosti «Lesnoe hozjajstvo»,* 264 p. (2011)

9. G.T. Selyaninov, Gidrometizdat, *In book: Zasuhi v SSSR, ih proishozhdenie, povtorjaemost i vlijanie na urozhaj*. pp. 5–30 (1958)
10. B.-C. Gao, Remote sens. Environ. **58**. pp.257-266 (1996)
11. A. S. Cherepanov, Geomatika. **2**. pp. 98–102 (2011) ISSN: 2410-6879
12. S.A. Bartalev, V.A. Egorov, V.O. Zharko, E.A. Lupjan, D.E. Plotnikov, S.A. Hvostikov, N.V. Shabanov. *Sputnikovoe kartografirovanie rastitelnogo pokrova Rossii*, 208 p. (2016)
13. N.D. Yakimov, E.I. Ponomarev, E3S Web of Conferences, **149**, 03008 (2020) https://doi.org/10.1051/e3sconf/202014903008
14. T.V. Ponomareva, K.Y. Litvintsev, K.A. Finnikov, N.D. Yakimov, A.V.Sentyabov, E.I. Ponomarev, Forests , **12**, p.994. (2021) https://doi.org/10.3390/f12080994

Artificial neural network technology for lips reading

Anna Pyataeva[1]* and *Anton* Dzyuba[1]

[1]Siberian Federal University, Institute of Space and Information Technology, 26, Kirensky st., Krasnoyarsk, 660074, Russian Federation

Abstract. The paper presents the use of neural networks for the task of automated speech reading by lips articulation. Speech recognition is performed in two stages. First, a face search is performed and the lips area is selected in a separate frame of the video sequence using Haar features. Then the sequence of frames goes to the input of deep learning convolutional and recurrent neural networks for speech viseme recognition. Experimental studies were carried out using independently obtained videos with Russian-speaking speakers.

1 Introduction

Visual speech recognition is a most important task when you communicating with hearing impaired people. According to the World Health Organization [1], as of 2020, more than 5% of all people in the world (360 million people) have serious hearing impairment problems. Hearing aids cannot fully solve the problem of these people because of hearing loss may be associated with neuropathology. There are diseases in which people lose the ability to make sounds. In such situations, speech recognition by lips articulation is one way to keep these people communication ability.

Speech recognition by articulation is based on the detection of visemes [2], i.e. mimic realizations of phonemes. There are 42 phonemes in Russian. 6 of them are vowels (а, и, о, у, ы, э) and 36 consonants (б, б', в, в', г, г', д, д', ж, з, з', э(й), к, к', л, л', м', м, н, н', п, п', р, р', с, с', т, т', ф, ф', х, х', ц, ч, ш, щ). Visemes and phonemes do not have a one-to-one correspondence. There are situations, when several phonemes correspond to one viseme and look the same on the speaking person face. The reason for this phenomenon is in fact that phonemes are reproduced inside the mouth or throat.

Traditionally, automatic lips reading (ALR) systems relied on the visual features extraction using hidden Markov models [4]. Currently, deep learning neural networks are used to solve the problems of automated recognition of various human actions (including lip reading) [3]. Speech recognition systems by articulation are used for dictating messages on smartphones [5], visual recognition of silent passwords [6], transcribing silent films [7], synthesizing the voice of people with speech impairments based on the movements of their lips [8], and developing systems for tracking the lips of the interlocutor for people with hearing impairments [9] and in other areas.

*Corresponding author: anna4u@list.ru

2 Algorithm for recognizing human speech by articulation

Lips speech recognition consists of two stages: face detection with lips highlighting and viseme recognition.

2.1 Face detection and lips area highlighting

To detect faces we used the OpenCV computer vision library [10] and the NumPy library [11]. The algorithm for face detection and lips area selection consists of the following steps.

1. Each frame of the video sequence is converted to grayscale since color does not play a significant role in detecting a face on the scene, but its presence negatively affects the execution speed of the algorithm.
2. Next step is detected a face in the examined frame using the Haar features [12] and then in the found area of the face algorithm is detected for lips.
3. The area with the detected lips is enlarged by 15 pixels on each side to prevent the lips from being cropped.
4. The lip areas obtained for each video sequence are stored in a NumPy data.

The resulting sequence of frames with images of lips is transmitted to a convolutional neural network for further processing.

2.2 Speech phoneme recognition by articulation

To recognize phonemes of speech by articulation, two neural networks are jointly used (Fig. 1): the MobileNet convolutional neural network (CNN) [13] and the recurrent neural network of the LTSM architecture [14]. The MobileNet architecture was chosen as the transfer learning model due to its light weight. The MobileNet convolutional neural network is used to normalize images. At the end of the CNN model is a fully connected layer, which output is used as input to the LSTM layer, so the ReLu function is chosen as the CNN activation function. The fully connected CNN layer to bring the data into the required dimension, has 1024, 128 neurons. After converting the videos using CNN, the video data looks like (586, 20, 128, 128, 3), where 586 is the total number of pronunciations for all tags, 20 is the number of frames for speaking, 128 and 128 are the width and height of the image, 3 is the number of color channels of the video image.

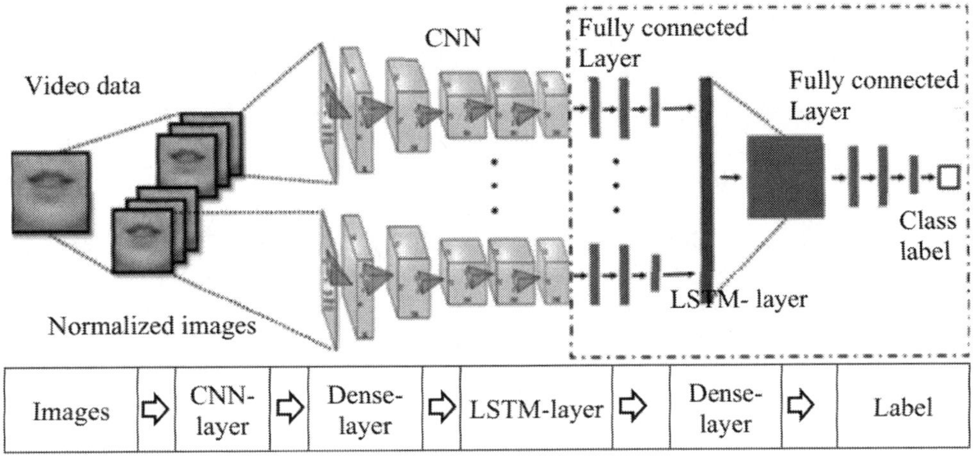

Fig. 1. Neural network architecture for speech recognition by articulation.

The LSTM-layer is used the TimeDistribute function supported by Keras. The TimeDistribute function takes ordered data and encodes its content. The CNN model is used as the object of coding. Combining CNN and LSTM models performed by using the Model function. Adam which has good performance was used as an optimizer. Since the model will perform the task of classification, the cross-entropy is chosen as the loss function.

The software implementation is performed with using the Keras deep learning library [15], which supports interaction with the Theano and Tensorflow packages. Keras functionality is very wide and has a convenient API, and its low speed is compensated by CUDA support. Keras is used to design, deploy and train a neural network, and to recognize speech visemes from a set of articulation frames.

3 Experimental research

Experimental studies used a dataset containing 768 different statements uttered by different speakers. The videos were obtained independently, as a dataset of Russian-speaking speakers was required. The utterances are labelled with the same labels as the training dataset. The training selection was 80%, the test selection was 20% of the total sample selection. The quality of the algorithm was evaluated using the "recognition accuracy" indicator. Test words were: «бегу» (in English means run), «пила» (in English means saw), «милый» in English means sweet), «усы» (in English means moustache), «вулкан» (in English means volcano), «банан» (in English means banana), «тонуть» (in English means drown). The best speech recognition accuracy by articulation was 93.7% for the word «банан» (banana) and the average accuracy was 68%, which suggests that the algorithm needs additional modification. Equations should be centred and should be numbered with the number on the right-hand side.

References

1. World Health Organization. https://www.who.int/
2. Speech Recognition: Watch what you say. https://www.econimist.com/ecience-and-technology/2015/01/23/watch-what-you-say
3. H. McGurk, J. MacDonald. Nature, 264, 7-18 (1976)
4. G. A. Fink. Berlin Heidelberg: Springer-Verlag. (2008)
5. A. Gabbay, T. Ephrat, S. Halperin, S. Peleg, *Seeing through noise: Speaker separation and enhancement using visually-derived speech,* in Proceedings of the International Workshop on Computer Vision for Audio-Visual Media, arXiv:1708.06767v1 (2017)
6. F. S. Lesani, F. F. Ghazvini, R. Dianat, *Mobile phone security using automatic lip reading* in Proceedings of the 9th International Conference on e-Commerce in Developing Countries: With focus on e-Business (ECDC), 16 April 2015, Isfahan, Iran (2015)
7. Y. M. Assael, B. Shillingford, S. Whiteson, N. De Freitas, *Lipnet: Sentence-level lipreading*, in Proceedings of the ICLR 2017 : 5th International Conference on Learning Representations, Toulon, France (2017)
8. A. Ephrat, T. Halperin, S. Peleg, *Improved speech reconstruction from silent video,* in Proceedings of the IEEE International Conference on Computer Vision Workshop (ICCVW), 22-29 October, 2017, Venice, Italy (2017)
9. A. Britto Mattos, D. A. Mattos, *Multi-view mouth renderization for assisting lip-reading,* in Proceedings of the Conference the Internet of Accessible Things. (2018)
10. NumPy. https://numpy.org

11. OpenCV. https://opencv.org
12. P. Viola, M. J.Jones, *Rapid Object Detection using a Boosted Cascade of Simple Features,* in Proceedings of the IEEE Conf. on Computer Vision and Pattern Recognition, Kauai, Hawaii (2001)
13. S. Hasim, W. Andrew, arXiv:1402.1128 (2014)
14. K. Yamaguchi, K. Sakamoto, T. Akabane, Y. Fujimoto *A Neural Network for Speaker-Independent Isolated Word Recognition,* in Proceedings of the First International Conference on Spoken Language Processing, 18-22 November, Kobe, Japan
15. Keras. https://keras.io

Using deep learning algorithms for texture segmentation of ultra-high resolution satellite images

Dmitry Rusin[1]*, *Anna* Alehina[1], *Anastasia* Safonova[1], and *Egor* Dmitriev[2]

[1] Deep Learning Laboratory, Siberian Federal University, 660041 Krasnoyarsk, Russia
[2] Marchuk Institute of Numerical Mathematics of the Russian Academy of Sciences, 119991 Moscow, Russia

Abstract. This paper presents the results of textural segmentation of satellite images with spatial resolution <1 m using U-Net convolutional neural networks. To conduct numerical experiments, a panchromatic image of the WorldView-2 test site on the territory of the Bronnitsky Forestry (Moscow region) used. The possibilities of automating the selection of neural network parameters based on genetic algorithms investigated. The proposed method makes it possible to effectively segment the main types of natural and man-made objects, as well as to distinguish structural classes of woodlands.

1 Method

In this paper, the task was set to create a machine-learning model based on convolutional neural networks. This model should automatically segment natural and technogenic objects on panchromatic satellite images of ultra-high spatial resolution by textural features. The *Xception* model used in the paper [1] with the ability to change a number of parameters to improve accuracy. Architecture *Xception* is a convolutional neural network that is capable of working with a small amount of training data for segmentation tasks. The generalized *U-Net* architecture [2] shown in Figure 1.

* Corresponding author: drusin@sfu-kras.ru

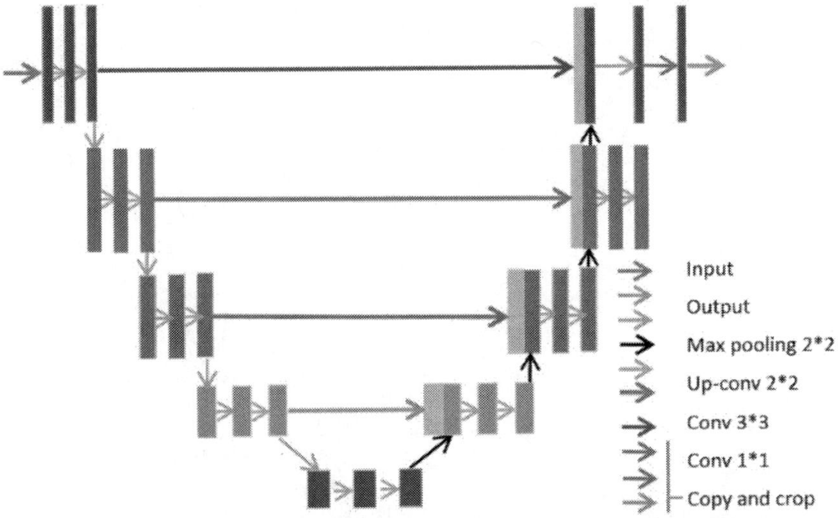

Fig. 1. Generalized U-Net architecture (figure from https://arxiv.org/abs/1505.04597).

Automated parameter setting performed using a genetic algorithm (GA). The algorithm based on the output quality assessment of the U-Net model. Calculations performed to assess the effectiveness of automated parameter tuning. The applied model trained in two ways, with the use of GA and without it. To compare the quality of the models, it decided to use cross-qualification [3] for evaluation on independent data. Cross-qualification applied only on the standard U-Net model without GA. The GA parameters presented in Table 1.

Table 1. GA parameters.

Name	Meaning
Mutation coefficient	0.1
Population	20
Number of iterations	50
Crossing	3

2 Data

The data set consists of satellite images of the Bronnitsky forestry of the Moscow region, the total image has a size of 10140 x 10120. First, this image processed: it divided into 64 parts for subsequent data markup.

Fig. 2. Splitting the source image into frames.

Splitting the snapshot allows you to reduce the load during training, as well as mark up the data in more detail. The final size of each frame is 1267x1265.

In this task, six classes were allocated: mixed forest, field, city, forest belt, ordered forest, water.

To assess the quality of the model, the mAP (4) metric used, based on the error matrix F-measure (3), it also represents a balance between two metrics Precision (1) and Recall (2). Categorical cross-entropy (5) used as the loss function of the model.

$$Precision = \frac{TP}{TP+FP} \quad (1)$$

$$Recall = \frac{TP}{TP+FN} \quad (2)$$

$$F\_score = 2 * \frac{Precision * Recall}{Precision + Recall} \quad (3)$$

$$mAP = \sum_{i=1}^{N} AP_i = \frac{1}{N} \sum_{Recall} \mathrm{Pr}ecision(\mathrm{Re}call_i) \quad (4)$$

$$Loss = - \sum_{i=1}^{Output\,size} y_i * \log y_i \quad (5)$$

where, y_i is the scalar value at the output of the model, output size is the number of scalar values in the output of the model.

An example of the markup of one of the frames shown in Figure 3.

Fig. 3. Example of a marked-up snapshot.

3 Results

The neural network architecture (NN) trained using a free Google Colab cloud-computing machine [4]. The main results of the experiment on test data presented in Table 2.

Table 2. Results of the work

Name of the metric	Unet	Unet+GA	Cross-qualification (folds = 5)
F-measure	75.74	77.86	82.37
Loss	22.12	16.75	13.00
mAP	71.54	73.56	76.49
Time spent on training (1 iteration, min)	*25*	55	130

The results of the work of U-Net and U-Net+GA display an average of 15 launches each. As can be seen from the table, the U-Net model copes with the processing of satellite images in the task of segmentation. For higher results, a more precise network setup needed, since even changing several parameters allowed to increase the accuracy of the model. In addition, the model is sensitive to the amount of input data. This seen based on the results of cross-validation. In the future, it is planned to test these models on new sites.

The paper was carried out with the financial support of the RFBR, projects No. 20-07-00370 "Fundamental problems of increasing the information content of data processing of optoelectronic aerospace devices of high spatial and spectral resolution" and No. 19-01-00215 "Investigation of the operational capabilities of hyperspectral remote sensing technologies for solving regional problems using existing and promising space-based GSCs".

References

1. F. Chollet, Xception: Deep Learning with Depthwise Separable Convolutions, arXiv:1610.02357v3 [cs.CV], 2017
2. O. Ronneberger, U-Net: Convolutional Networks for Biomedical Image Segmentation, arXiv:1505.04597, 2015
3. F. Mosteller and J.W. Tukey Data analysis, including statistics. In Handbook of Social Psychology. Addison-Wesley, Reading, MA, **46**, 1968.
4. Google Colab [Electronic resource] - Access mode: https://colab.research.google.com

Development of compression algorithms for hyperspectral aerospace images based on discrete orthogonal transformations

Assiya Sarinova

Kazakh Agrotechnical University named after S. Seifullin, 010000, Zhenis avenue 62, Kazakhstan

Annotation. The paper describes the development of compression algorithms for hyperspectral aerospace images based on discrete orthogonal transformations for the purpose of subsequent compression in Earth remote sensing systems. As compression algorithms necessary to reduce the amount of transmitted information, it is proposed to use the developed compression methods based on Walsh-Hadamard transformations and discrete-cosine transformation.
The paper considers a methodology for developing lossy and high-quality compression algorithms during recovery, taking into account which an adaptive algorithm for compressing hyperspectral AI and the generated quantization table has been developed. The conducted studies have shown that the proposed lossy algorithms have sufficient efficiency for use and can be applied when transmitting hyperspectral remote sensing data in conditions of limited buffer memory capacity and bandwidth of the communication channel.

Introduction

Hyperspectral AI remote sensing is necessary for monitoring natural resources and the consequences of emergencies, etc. Currently, the development of software systems for lossy data compression is an urgent task. In solving this problem, there are various areas of research in which research is actively being conducted in the field of developing compression algorithms [1-13].

Algorithms for processing hyperspectral AI with losses based on discrete transformations have been developed. The sequence of stages is as follows.

1. The original transformation of the data structure based on the original hyperspectral AI, storing the values of the coefficients, based on the discrete-cosine transformation with the generated quantization table.

2. Transformation of the data structure based on the original hyperspectral AI, coefficient values, based on Walsh-Hadamard three levels.

3. Transformation of the obtained data structures based on steps 1-2 by means of the generated coefficient quantization table.

4. Using standard criteria for the quality of restored images.

5. Compression of the obtained structures of stage 4 by one of the standard entropy algorithms.

6. Experimental study of conversion algorithms by compression ratio and quality of recovered data.

1 Walsh - Hadamard transformation of three levels

Consider an example of the Walsh-Hadamard transformation (WHT) for a fragment of hyperspectral AI.

Let the fragment AI represent a matrix consisting of *m* rows, *n* columns and *k* channels: **I**[*m*, *n*, k] = **I**[10, 10, 10].

The WHT transform is filters that divide the image into low-frequency and high-frequency components. To get the original image, you just need to combine these components again. The direct WHT transformation is represented in matrix form H_{wt2}, a fragment of AI is taken, т.е. $H_{wt2} \bullet I[m,n,k] = H_{wt2}'[CK]$, где where SC – spectral component of the matrix **I**[*m*, *n*, k].

$$\begin{pmatrix} \frac{1}{2} & \frac{1}{2} & 0 & 0 & 0 & 0 & 0 & 0 & 0 & 0 \\ \frac{1}{2} & -\frac{1}{2} & 0 & 0 & 0 & 0 & 0 & 0 & 0 & 0 \\ 0 & 0 & \frac{1}{2} & \frac{1}{2} & 0 & 0 & 0 & 0 & 0 & 0 \\ 0 & 0 & \frac{1}{2} & -\frac{1}{2} & 0 & 0 & 0 & 0 & 0 & 0 \\ 0 & 0 & 0 & 0 & \frac{1}{2} & \frac{1}{2} & 0 & 0 & 0 & 0 \\ 0 & 0 & 0 & 0 & \frac{1}{2} & -\frac{1}{2} & 0 & 0 & 0 & 0 \\ 0 & 0 & 0 & 0 & 0 & 0 & \frac{1}{2} & \frac{1}{2} & 0 & 0 \\ 0 & 0 & 0 & 0 & 0 & 0 & \frac{1}{2} & -\frac{1}{2} & 0 & 0 \\ 0 & 0 & 0 & 0 & 0 & 0 & 0 & 0 & \frac{1}{2} & \frac{1}{2} \\ 0 & 0 & 0 & 0 & 0 & 0 & 0 & 0 & \frac{1}{2} & -\frac{1}{2} \end{pmatrix} * \begin{bmatrix} 123 \\ 105 \\ 121 \\ 103 \\ 118 \\ 100 \\ 123 \\ 123 \\ 122 \\ 104 \end{bmatrix} = \begin{bmatrix} \frac{123}{2}+\frac{105}{2} \\ \frac{123}{2}-\frac{105}{2} \\ \frac{121}{2}+\frac{103}{2} \\ \frac{121}{2}-\frac{103}{2} \\ \frac{118}{2}+\frac{100}{2} \\ \frac{118}{2}-\frac{100}{2} \\ \frac{123}{2}+\frac{123}{2} \\ \frac{123}{2}-\frac{123}{2} \\ \frac{122}{2}+\frac{104}{2} \\ \frac{122}{2}-\frac{104}{2} \end{bmatrix} = \begin{bmatrix} \frac{228}{2} \\ \frac{18}{2} \\ \frac{224}{2} \\ \frac{18}{2} \\ \frac{218}{2} \\ \frac{18}{2} \\ \frac{246}{2} \\ 0 \\ \frac{226}{2} \\ \frac{18}{2} \end{bmatrix}$$

As a result, after such a transformation, we obtain the coefficients of the low-frequency and high-frequency components SC = 114, 9, 112, 9, 114, 9, 123, 0, 123, 9. During quantization, high-frequency coefficients (close to zero and negative values) are rounded to zero. It should be noted that the level matrices H_{wt4} и H_{wt8} are calculated in the same way as and H_{wt2}.

At the stage of restoring the original image channels, the SC is decoded. An example of the reverse transformation is presented as follows:

$$\begin{pmatrix} \frac{1}{2} & \frac{1}{2} & 0 & 0 & 0 & 0 & 0 & 0 & 0 & 0 \\ \frac{1}{2} & -\frac{1}{2} & 0 & 0 & 0 & 0 & 0 & 0 & 0 & 0 \\ 0 & 0 & \frac{1}{2} & \frac{1}{2} & 0 & 0 & 0 & 0 & 0 & 0 \\ 0 & 0 & \frac{1}{2} & -\frac{1}{2} & 0 & 0 & 0 & 0 & 0 & 0 \\ 0 & 0 & 0 & 0 & \frac{1}{2} & \frac{1}{2} & 0 & 0 & 0 & 0 \\ 0 & 0 & 0 & 0 & \frac{1}{2} & -\frac{1}{2} & 0 & 0 & 0 & 0 \\ 0 & 0 & 0 & 0 & 0 & 0 & \frac{1}{2} & \frac{1}{2} & 0 & 0 \\ 0 & 0 & 0 & 0 & 0 & 0 & \frac{1}{2} & -\frac{1}{2} & 0 & 0 \\ 0 & 0 & 0 & 0 & 0 & 0 & 0 & 0 & \frac{1}{2} & \frac{1}{2} \\ 0 & 0 & 0 & 0 & 0 & 0 & 0 & 0 & \frac{1}{2} & -\frac{1}{2} \end{pmatrix} * \begin{bmatrix} \frac{228}{2} \\ -\frac{18}{2} \\ \frac{224}{2} \\ \frac{18}{2} \\ \frac{218}{2} \\ \frac{18}{2} \\ \frac{246}{2} \\ 0 \\ \frac{226}{2} \\ \frac{18}{2} \end{bmatrix} \begin{bmatrix} \frac{228}{2}+\frac{18}{2} \\ \frac{228}{2}-\frac{18}{2} \\ \frac{224}{2}+\frac{18}{2} \\ \frac{224}{2}-\frac{18}{2} \\ \frac{218}{2}+\frac{18}{2} \\ \frac{218}{2}-\frac{18}{2} \\ \frac{246}{2}+0 \\ 0 \\ \frac{226}{2}+\frac{18}{2} \\ \frac{226}{2}-\frac{18}{2} \end{bmatrix} = \begin{pmatrix} \frac{246}{2} \\ \frac{210}{2} \\ \frac{242}{2} \\ \frac{206}{2} \\ \frac{236}{2} \\ \frac{200}{2} \\ \frac{246}{2} \\ 0 \\ \frac{246}{2} \\ \frac{208}{2} \end{pmatrix}$$

There $H^T_{wt2} \bullet H_{wt2}'[CK] = I[m,n,k]$, where H^T_{wt2} - inverse wavelet transform WHT, $H_{wt2}'[CK]$ - spectral component.

The advantage of the Walsh-Hadamard transformation is an increased ability to detect low-frequency components due to the separation of hyperspectral AI channel regions into sublevels and the generated quantization coefficient.

2 Discrete-cosine transformation with generation of quantization tables

The discrete cosine transform (DCT) is a transformation whose linear combination consists of known basis vectors weighted with n coefficients leading to the original vector.

Let's consider step-by-step transformation of hyperspectral AI channels using DCT.

1. Divide it into s blocks of pixels of size nxn (usually 8 x 8).
2. Apply PREP to each block, represent each block as a linear combination of 64 basic blocks.
3. All ***s*** vectors $W^{(i)}$ (i= 1,2….s)
4. Generation of the quantization table.

At the first stage, as a result of the DCT conversion, two filters were built – high-frequency and low-frequency. For low-frequency and high-frequency filters, we will introduce some designations: DC is the operator for the low-frequency filter; AC is the operator for the high-frequency filter. After the DCT transformation, the fragment of hyperspectral AI has the following values (Fig. 1):

40008	204	708	-320	268	-139	-51	51
799	-176	654	-84	60	89	-57	114
188	-64	-79	102	-64	137	-55	26
326	29	157	-119	-47	79	-56	-1
97	-94	173	25	44	129	-26	-63
33	128	-9	118	-91	14	-96	-12
-207	41	-198	118	-28	34	1	-18
13	-1	-3	-14	-2	-35	-20	-11

Fig.1. The coefficients obtained after the DCT conversion

It can be seen from Fig.1 that the high values of the conversion coefficients are concentrated in the upper left corner and only a small number of low-frequency coefficients prevail over the rest. This allows you to reduce the values at the next stages of compression.

The second stage is the process of generating quantization tables. Quantization is performed in this way:

1. The arithmetic mean $\mathbf{I}_{av}$ was calculated from all the values of the file.

2. To select the numerical quantization index Quant, the number for quantization was calculated, $Quant$ = quantValue × $\mathbf{I}_{av}$/100. For example, the numerical quantization index: $\mathbf{I}_{av}$ * 95 % = 285 = $Quant$. At the same time, the DC coefficients remain intact. For each hyperspectral AI channel, its own quantization table is generated.

For the above fragment of hyperspectral AI, the matrix after quantization is shown (Fig.2). The quantization process is key in the compression process, the advantage of representation in the frequency domain lies in the visual quality of the reconstructed images. As a result, after quantization, most of the coefficients are zero.

40008	0	708	-320	268	0	0	0
799	0	654	0	0	0	0	0
0	0	0	0	0	0	0	0
326	0	0	0	0	0	0	0
0	0	0	0	0	0	0	0
0	0	0	0	0	0	0	0
0	0	0	0	0	0	0	0
0	0	0	0	0	0	0	0

Fig. 2. Quantization of the DCT coefficients

As a result, most of the zeros will be placed at the end of the data compression stream. This stream with many consecutive zeros at the end of the block is optimized to achieve high compression in the entropy coding of the adaptive Huffman algorithm.

2 Experiments of the developed algorithms of lossy transformations

To determine the effectiveness of the proposed adaptive transformation algorithms in terms of the degree of compression and quality of the recovered images, as well as the limits of its applicability, a number of experiments were conducted on hyperspectral AI (*Aviris* remote sensing system), (Fig.3). The experiments were performed on a PC with an *IntelCore i5* processor 2.5 GHz and 4 GB RAM running *Windows 8.1* operating system.

Indicators of compression degrees in terms of the number of channels of orthogonal Walsh–Hadamard transformations, DCP and JPEG 2000 *Lossy* are shown in Fig. 3. It is shown that *DCT* prevails over the Walsh-Hadamard and JPEG *Lossy* transformations in the compression ratio with high quality of the restored images. As can be seen from Fig.3, the indicators of the degrees of compression of the DCT with losses are superior in the degree of compression of the Walsh-Hadamard transformation and the *JPEGLossy* compressor.

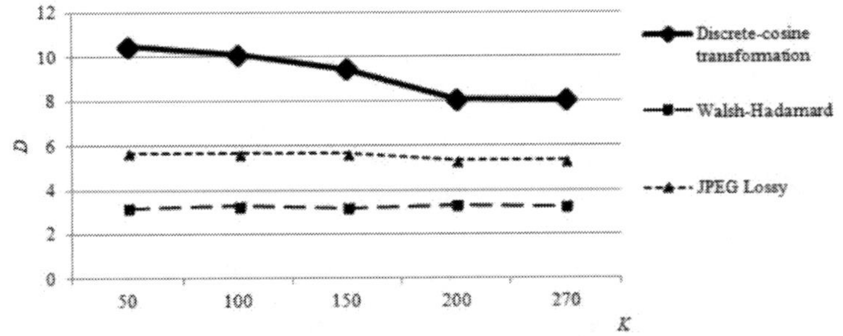

Fig. 3. Dependencies of R on D

Indicators of the quality metrics of the restored images were determined using *PSNR* and *SKO*. The degree of distortion compares the ratio between compression and distortion in lossy algorithms. The score is defined as the average of the number of bits needed to represent each pixel. Measured in bits per pixel (bpp - bits per pixel). Distortion is usually measured using *PSNR*. Figure 4.28 shows the dependence of *PSNR* on the compression ratio *D*. The following Figure 5 shows the dependence of *PSNR* on *D* for *DCT* and the Walsh-Hadamard transformation.

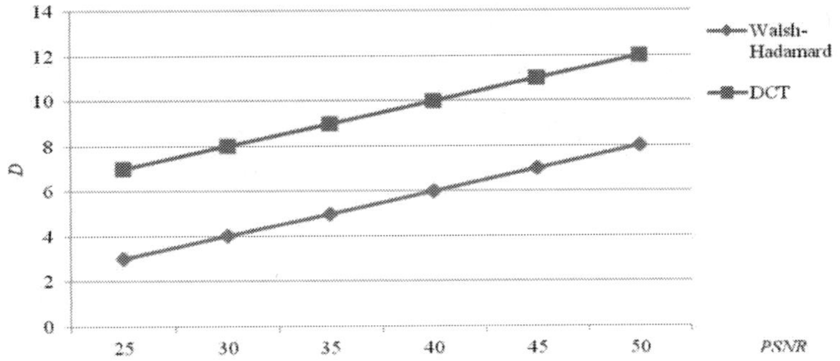

Fig. 4. PSNR dependence on For Walsh-Hadamard and DCT

Conclusion

Based on the above results of the compression experiment based on orthogonal and wavelet transformations of hyperspectral AI with losses, it can be concluded that the developed mathematical and compression software, to some extent occupies a leading position in the degree of compression and quality of the restored images, depending on the selected preparatory processing, significantly demonstrates and surpasses analogues on all test sets. Based on the conducted research , the following conclusions should be drawn:

- the development of algorithms based on discrete transformations makes it possible to increase the compression ratio to (R=12);
- the proposed lossy approach is defined in adaptive transformations based on the Walsh-Hadamard transformation, discrete-cosine transformation and the generated quantization table and subsequent adaptive coding.

Литература:

1. D. S. Sujithra, T. Manickam, D. S. Sudheer. Compression of hyperspectral image using discrete wavelet transform and walsh hadamard transform. International Journal of Advanced Research in Electronics and Communication Engineering. Vol. 2(3), 314 – 319 pp. 2013.
2. P. Keerthana, A. Sivasankar. The impact of lossy compression on hyperspectral data adaptive spectral unmixing and PCA classification. International Journal of Science and Modern Engineering, Vol.1(7), 35-37 pp. 2013.
3. Poonam and R.S. Chauhan. Compression and Classification of Hyperspectral Images using an Algorithm based on DWT and NTD. Advance in Electronic and Electric Engineering. Vol.3(4). 447-456 pp. 2013.
4. L. Santos, E. Magli, R. and et. Lossy hyperspectral image compression on a graphics processing unit: parallelization strategy and performance evaluation. Journal of Applied Remote Sensing. Vol. 7, 074955-1-15 pp. 2013.
5. S.M. Ramesh, P. Bharat, J. Anand, J. AnbuSelvan. Analysis of Lossy Hyperspectral Image Compression Techniques. International Journal of Computer Science and Mobile Computing, Vol.3(2), 302-307 pp. 2014.
6. A. Sarinova, A. Zamyatin. Hyperspectral regression lossless compression algorithm of aerospace images. E3S Web *Conferences* **149**, 02003 (2020).
7. S. Kudubayeva, N. Amangeldy, A. Sundetbayeva and A. Sarinova. ACM International Conference Proceeding Series **8**, (2019).
8. A. Sarinova, A. Zamyatin and P. Cabral DYNA (Colombia) **82** (190), 166-172 (2015).
9. A. V. Neftissov, O. A. Andreyeva, Sarinova A. Zh., Investigation of the properties of reed switches in devices for resource-saving relay protection of the electrical part of power plants. *AIP Conference Proceedings 2337,* No. 030010, 2021.
10. A. Zh. Sarinova, A. V. Drobinsky, L. N. Kirichenko, Using and applying computer technologies for studying transient processes in electrical circuits. *AIP Conference Proceedings 2337,* No. 030013, 2021.
11. Sarinova A., Zamjatin A., Methodology for Developing Algorithms for Compressing Hyperspectral Aerospace Images used on Board Spacecraft, *E3S Web of Conferences* 223, *02007,* 2020.
12. S. Abimuldina, A. Sarinova, L. Sarlybayeva and N. Akhmetova. DYNA (Colombia) 84 (202), 289-294 (2017).

13. A. Sarinova, A. Zamyatin and P. Cabral. Lossless compression of hyperspectral images with pre-byte processing and intra-bands correlation. DYNA (Colombia) 82 (190), 166-172 (2015).

Digital processing of space images using global filtering in the frequency area

Valeriy Tutatchikov[1*], *Mikhail* Noskov [1]

[1] Siberian Federal University, School of Space and Information Technology, 660074, Krasnoyarsk, Russian Federation

Abstract. In modern systems, remote sensing widespread two-dimensional fast Fourier transform (FFT) for processing satellite images and the subsequent image filtering. In this paper we consider the possibility of using combinations of high and low frequency filters to improve global images terrestrial surface.

1 Introduction

Currently, remote sensing of the Earth is closely related to digital image processing, since the obtained aerospace images are presented in digital form in the form of raster images. To improve images, usually used are changes in brightness and contrast, spatial filtering, Fourier transform and subsequent frequency filtering [1]. Consider in this paper the possibility of using a combination of the results of applying high and low pass filters to improve the image.

2 Method description

Let us consider as the initial data the image of Krasnoyarsk and its environs from the LandSat-8 satellite dated April 7, 2016 [2] (Figure 1). For the convenience of calculations, let us set the size of the investigated area as a power of two: 512 * 512 pixels. This image can be represented as a function $f(x,y)$ of the brightness of a pixel with coordinates x and y, taking integer values 0-255 for each red, green, blue channel in the RGB color model. In [1], it is proposed to multiply all the elements of the image $f(x,y)$ by $(-1)^{x+y}$, and then calculate the discrete Fourier transform of this function [3]. In this case, we get a centered Fourier transform $F(u,v)$ in the frequency domain, in which high frequencies (responsible for image contours and sharp transitions) will be in the center of the spectrum, and low frequencies (responsible for image smoothing and color saturation) will be around the edges.

* Corresponding author: vtutatchikov@mail.ru

Fig. 1. Satellite image of Krasnoyarsk surface

We apply a global high-pass filter to the obtained function *F(u,v)*, for example, a Gaussian high-pass filter (HPF):

$$H(u,v)=1-\exp\left(\frac{-D^2(u,v)}{2\cdot D_0^2}\right) \quad (1)$$

where *D(u,v)* is the distance from the center of the Fourier transform to the point with coordinates *(u,v)*, D_0 is the cutoff frequency. The Gaussian HPF graph is shown in Figure 2.

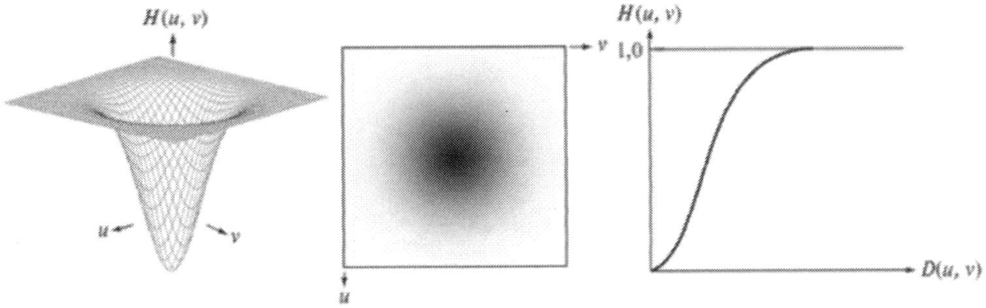

Fig. 2. Graph of a Gaussian HPF function on the left, a grayscale image of the filter in the center, radial filter profiles for different values of D_0 on the right

As a result of filtering, the high frequencies will remain unchanged, and the low frequencies outside the radius D_0 will be zeroed. After that, we calculate the inverse two-dimensional Fourier transform, multiply all the samples of the resulting function by $(-1)^{x+y}$ to cancel centering, and we get the processed image $f_1(x,y)$ (Figure 3 on the left).

Fig. 3. The result of applying a Gaussian LPF on the left, a Gaussian HPF with $D_0 = 1$ on the right

Then to the function $F(u,v)$ a global low-pass filter, for example, a Gaussian low-pass filter (LPF):

$$H(u,v)=\exp\left(\frac{-D^2(u,v)}{2\cdot D_0^2}\right) \quad (2)$$

where $D(u,v)$ is the distance from the center of the Fourier transform to the point with coordinates (u,v), D_0 is the cutoff frequency. The Gaussian LPF graph is shown in Figure 4.

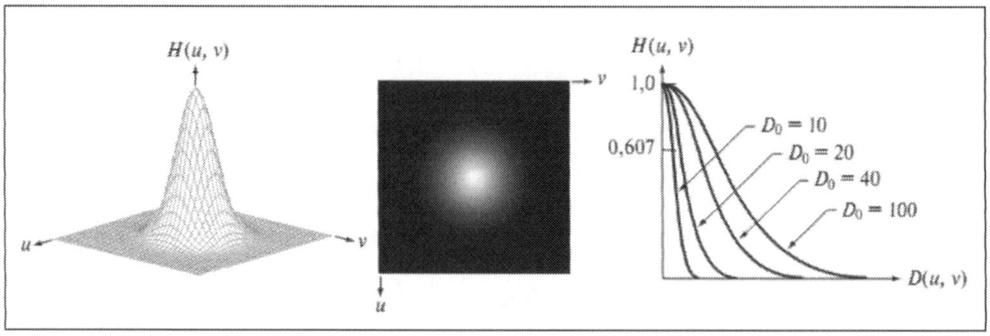

Fig. 4. Graph of a Gaussian LPF function on the left, a grayscale image of the filter in the center, radial filter profiles for different values of D_0 on the right

As a result of filtering, the low frequencies will remain unchanged, and the high frequencies outside the D_0 radius will be zeroed. After that, in the same way, we calculate the inverse two-dimensional Fourier transform, multiply all the samples of the resulting function by $(-1)^{x+y}$ to cancel centering, and we get the processed image $f_2(x,y)$ (Figure 3 on the right).

Then we perform the step of combining the resulting images. For each pixel of the new image $f_3(x,y)$, recalculate the brightness value using the formula

$$f_3(x,y)=a\cdot f_1(x,y)+b\cdot f_2(x,y) \quad (3)$$

where $a,b>=0, a+b=1$. In this case $a=b=0.5$.

3 Obtained results

The final filtering result is shown in Figure 5.

Fig. 5. Global high-pass filtering result

The contours of objects have become noticeably outlined in the image, the color saturation has increased.

4 Conclusion

On the example of a satellite image of Krasnoyarsk and its environs, a method of using a combination of global high- and low-frequency filtering to improve the image is shown. As a result of processing, image details become more visible and colors brighter.

Acknowledgments

This work was carried out as part of the state assignment of the Ministry of Science and Higher Education of the Russian Federation (scientific code FSRZ-2020-0011).

https://doi.org/10.1051/e3sconf/202133301012

References

1. R.C. Gonzalez and R.E. Woods. *Digital Image Processing,* 1104 (2012)
2. Snapshot of the Taseevsky district of the Krasnoyarsk Territory from 07.30.2020. Access mode: http://earthexplorer.usgs.gov/
3. M. Noskov, V. Tutatchikov, *Proceeding of 11th International Forum on Strategic Technology (IFOST-2016)*, 495-498 (2016)

Using of Google Earth Engine in monitoring systems

Elena Fedotova[1*], and *Anna* Gosteva[1,2]

[1]FRC KSC SB RAS, Krasnoyarsk, Russia
[2]SFU, Krasnoyarsk, Russia

Abstract. Google Earth Engine (GEE) cloud service is a powerful tool for environmental research. An example of using GEE to solve a typical research problem is shown. The following data extraction and analysis operations were used: filtering data from sets, constructing functions, building graphs, selecting data using vector and raster masks. GEE interface in the form of JavaScript code was used. Correlation between surface runoff and precipitation and snow depth in areas with forest dieback was analysed for Krasnoyarsk region in Russia ($r = 0.30$ for precipitation and $r = 0.57$ for snow depth).

1 Introduction

Cloud GIS offers powerful software and hardware for solving environmental monitoring and research tasks by implementing geoinformatics methods: open access geographic servers with electronic map generation and implementation of multivariate analysis algorithms. Cloud GIS allows to receive data in real or near-real time and display them directly on the user's system. The Google Earth Engine platform [1] is unique as an integrated platform for empowering not only remote sensing professionals, but also a much wider audience that lacks the technical capabilities required to use traditional supercomputers or large-scale cloud computing resources [2].

GEE includes an interactive application server working with open data catalogue, computing environment in the form of a code editor (integrated development environment (IDE) for writing and running scripts), geospatial API (client libraries provide Python and JavaScript wrappers for web APIs in REST architecture).

The data catalogue contains a large repository of publicly available geospatial datasets, including images from a variety of satellite and aerial systems in both optical and microwave wavelengths, environmental variables, weather and climate forecasts, land cover maps, topographic and socio-economic datasets. All of this data is pre-processed into a ready-to-use form that provides efficient access and removes many barriers associated with data management. Operators are implemented in a parallel processing system that automatically splits and distributes computations, providing high-throughput analysis capabilities. Users access the API through a thin client library or through an interactive web development environment built on top of this client library [2].

* Corresponding author: elfed@ksc.krasn.ru

Data collections have descriptions and sample JavaScript codes for opening and rendering data in GEE. GEE client libraries are proxy objects for images (ee.Image), collections (ee.ImageCollection, ee.FeatureCollection), and other data types: numbers (ee.Number), strings (ee.String), vector objects with geometry (ee.Geometry), lists (ee.List). Images grouped by source, destination, etc. are combined into image collections (ImageCollection), that have the functionality to select data from the collection by filtering and sorting according to specified spatial, temporal, or other criteria. The API library allows performing operations on images using raster algebra and support of higher-order functions: map () and iterate (), that allow to apply arbitrary functions to image collections. The reduce () operator is used to calculate statistics on image collections, which, for example, can aggregate data using the sliding window method [2, 3]. Users can load their own data (both raster and vector type) using the library of operators provided by the Earth Engine API. Users actively use GEE to work with remotely sensed data. [4].

Figure 1 shows a working window, which displays a table of user's scripts, a part of the JavaScript code for data visualization, an image in a map window and a graph of parameter's values in console.

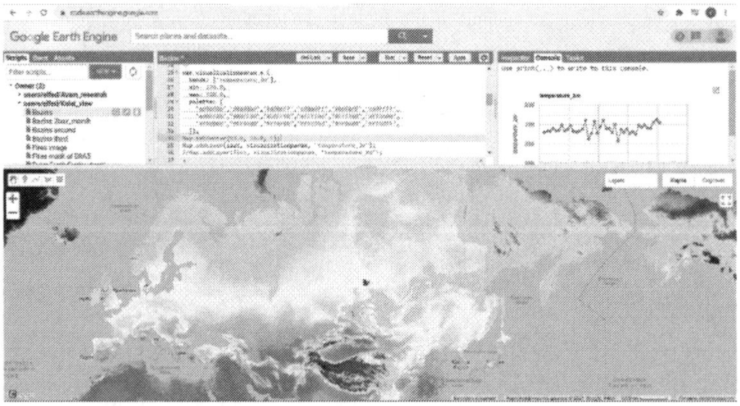

Fig.1. The Earth Engine interactive development environment

Thus, users have a set of analysis tools for spatial data visualization and processing. With GEE it is possible to create a very wide range of GIS analyses.

2 Data and methods

In order to develop a universal set of simple operations in GEE for solving monitoring tasks, a typical task of analysing the relationship between two environmental parameters - the dynamics of surface runoff and the dynamics of vegetation in a river basin - was set.
Surface flow dynamics in the areas where forest cover was destroyed in 2013 and 2016 was analysed. The regions were identified according to the Hansen forest cover change project [5]. These are the territories of the Kas and Sym basins, two left tributaries of the Yenisei River in Krasnoyarsk Region, Russia.

Hydrological and climatic data were selected from ERA5-Land Monthly Averaged [6]. The data cover the time interval from January 1981 to August 2021 (at this time, in general three months behind real time) and have a spatial resolution of 0.1 degrees. ERA5-Land is one of Climate Data Store (CDS) land surface dataset, from 1981 for monthly averaged values to present time (2-3 months in arrears), produced at higher resolution (9km) and forced by ERA5 atmospheric parameters with lapse rate correction, but with no additional data assimilation [7]. Reanalysis datasets are most suitable for spatiotemporally consistent environmental analysis [8]. The ERA5 Land dataset is actively used in climate change research [9-11].

From the parameter values (50 parameters) represented mean daily value for every month in year, five were used:

– surface runoff, is the depth the water would have if it were spread evenly over the grid box, m;

– total precipitation, accumulated liquid and frozen water, including rain and snow, that falls to the Earth's surface, m;

– total evaporation, accumulated amount of water that has evaporated from the Earth's surface, including a simplified representation of transpiration (from vegetation), into vapour in the air above, m;

– air temperature, Temperature of air at 2m above the surface of land, K;

– snow depth water equivalent (SDWE), depth of snow from the snow-covered area of a grid box, m of water equivalent [6].

The sequence of actions is here presented to solve the above task: data selection, their visualization, selection or calculating the required parameters, analysing their dynamics (plotting graphs, calculating statistics).

The data selection from the database and their visualization is performed using filtering operation. The characteristics of the data in the database and codes for this operations are presented in the GEE for all data collections. Five above parameters from the ERA5-Land database from 2006 to 2020 from March to November for each year were separated. JavaScript code for surface runoff is shown (fig.2).

```
// Create an ImageCollection from an Earth Engine Table.
var dataset = ee.ImageCollection("ECMWF/ERA5_LAND/MONTHLY")
                .filter(ee.Filter.date('2006-01-01', '2020-12-31'))
                .filter(ee.Filter.calendarRange(03, 11,'month'));
//select parameter for obteining
      var parameter = dataset.select('surface_runoff');
```

Fig. 2. JavaScript code of surface runoff filtering from 2006 to 2020 years for March-November

In the next step, it is necessary to extract characteristics for objects (by masks) from the filtered data: by vector objects - points, including the selection of pixels in the Inspector, along a line - profiles, by a polygon. To highlight the analysed area, it can be used raster masks from other raster datasets. In our task data from ERA5–Land DB were selected using the raster deforestation map [3] as a raster masks (fig.3).

```
var image = ee.Image("users/elfed/mask101");
var lossArea = image.select('b1');
var collection_ERA5 = parameter.map(function(img) {
                 return img.updateMask(lossArea);
               });
```

Fig. 3. JavaScript code for raster mask realization: mask101 – forest loss raster map, b1 – name of layer in mask101.

Values of five above mentioned parameters for 2006-2020 corresponding to this mask were exported from the GEE in csv format using print operation.

Reduce operation is used for per-pixel calculation of the statistic for every image in image collection. Collection of images is reduced to a single image. For basic statistics like min, max, mean, etc., ImageCollection has quick access methods like min (), max (), mean (), etc. Both built-in and user-created functions are used to calculate specific characteristics of the data, such as vegetation indices. The function is applied to every image in the collection. Function to calculate daily values of runoff, total precipitation, total evaporation in millimetres and monthly values was made (fig. 4).

```
// Create function (*30 - number of days in month, *1000 - mlm)
var mult30 = function(image) {
  var mult = ee.Image(0).expression(
    'runoff * 30*1000', {
      runoff: image.select('runoff'),
});
  return image.addBands(mult.rename('mult_mil_mon'));
};
print(Chart.image.series(parameter,      Kas,      ee.Reducer.mean(),
30).setOptions(options));
```

Fig. 4. JavaScript code for printing the graph of monthly mean values of surface runoff in millimetres in Kas river basin.

It is possible to output a graph with several values, for every polygon in the vector layer, for example to facilitate export in csv format (fig. 5).

Thus, for each step of data analysis JavaScript code has been created using GEE data descriptors and other auxiliary tools. By changing parameter names, time ranges, palettes for map visualization, equations in functions, the above scripts can be used for other data.

```
var Mothpoly = ee.FeatureCollection("users/elfed/Moth");
// Create a ImageCollection from an Earth Engine Table.
var dataset = ee.ImageCollection("ECMWF/ERA5_LAND/MONTHLY")
                .filter(ee.Filter.date('2006-01-01', '2020-12-31'))
                .filter(ee.Filter.calendarRange(1, 12,'month'));
var parameter = dataset.select('temperature_2m');
// Define the chart and print it to the console.
var chart =
    ui.Chart.image
        .seriesByRegion({
          imageCollection: dataset,
          band: 'temperature_2m',
          regions: Mothpoly,
          reducer: ee.Reducer.mean(),
          scale: 500,
          seriesProperty: 'Id',
          xProperty: 'system:time_start'})
        .setOptions({
          title: 'temperature_2m',
          hAxis: {title: 'Date', titleTextStyle: {italic: false, bold:
true}},
          vAxis: {
            title: 'temperature_2m',
            titleTextStyle: {italic: false, bold: true}},
          lineWidth: 5,
          colors: ['f0af07', '0f8755', '76b349'],});
print(chart);
```

Fig. 5. JavaScript code for printing the graph of mean air temperature for three polygons (2013, 2016, 2017 years of forest loss) in Moth vector layer.

3 Results and discussion

Extracted from ERA5–Land DB data were analysed in Microsoft Excel. Surface runoff anomalies were calculated, only for April – October months of each year.

The results are demonstrated in the fig. 6. A noticeable change (both a decrease and an increase) in the runoff value for the year following the destruction of forests is not observed for either 2013 or 2016. There is some correlation with precipitation in the same month (r = 0.30) and sum of SDWE in October-May (r = 0.57). The high level of May runoff in 2014 and 2015 in both areas occurs against the background of a high level of SDWE sum from October to May.

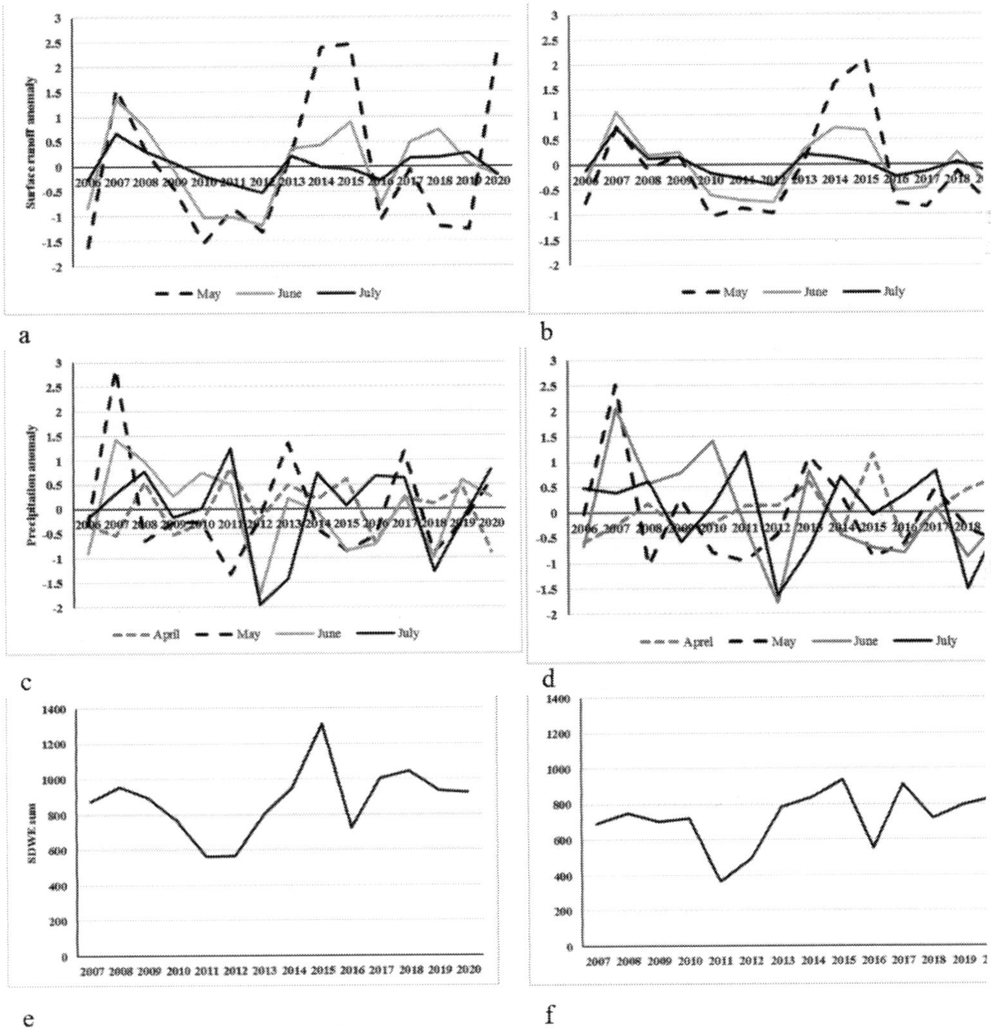

Fig. 6. Dynamics of surface runoff anomalies in areas of deforestation in 2013 and 2016 (a, b). Precipitation anomalies (c, d) and sum of SDWE for October – May (e, f). Surface runoff and precipitation – daily average values for a month.

Visual analysis of the 2015-2020 winter Landsat images was carried out in GEE without downloading to the user's computer. As can be seen in the winter satellite images (fig. 7), forest mortality results in a gradual change of the land cover over several years. Only

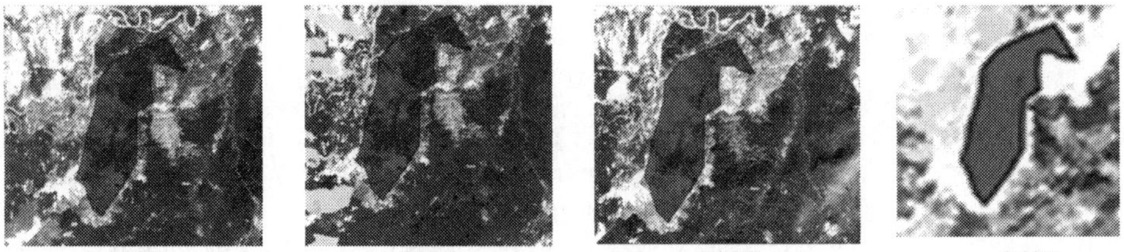

Fig. 7. Forest areas with damaged in 2013 year stands on the Landsat images of 2015-2020 years (true color).

in the 2020 image the site looks like treeless. This gradual change in the surface may also result in no abrupt change (increase) in surface runoff.

The research was funded by RFBR, Krasnoyarsk Territory and Krasnoyarsk Regional Fund of Science, project number 20-44-240007.

References

1. Google Earth Engine, https://earthengine.google.com/
2. N. Gorelick, M. Hancher, M. Dixon, S. Ilyushchenko, D. Thau, R. Moore, Rem. Sens. of Envir,. **202,** 18 – 27 (2017)
3. V. Lobanov, https: habr.com/ru/post/500020/ (2020)
4. A.A. Jamali, R.G. Kalkhajeh, T.O. Randhir, S. He, J. Environ. Manag., **302**, 113970 (2022)
5. M. C.Hansen, P. V. Potapov, R. Moore, M. Hancher, S. A. Turubanova, A. Tyukavina, D. Thau, S. V. Stehman, S. J. Goetz, T. R. Loveland, A. Kommareddy, A. Egorov, L. Chini, C. O. Justice, J. R. G. Townshend, Sci, **342,** 850 – 853 (2013)
6. ERA5-Land Monthly Averaged - ECMWF Climate Reanalysis, https://developers.google.com/earth-engine/datasets/catalog/ECMWF_ERA5_LAND_MONTHLY?hl=en#description
7. ERA5 dataset, https://confluence.ecmwf.int/display/CKB/ERA5
8. H. Zandler, T. Senftl, K.A. Vanselow, Sci, **10**, 22446 (2020)
9. S. Gleixner, T. Demissie, G.T. Diro, Atmosphere, **11**, 996 (2020)
10. A.K. Betts, D.Z. Chan, R.L. Desjardins, Front. Environ. Sci. (2019)
11. M.H. Pesci, F. Voges, N. Rüther, K. Förster, EGU General Assembly, https://doi.org/10.5194/egusphere-egu2020-2405 (2020)

The influence the horizontal structure of the forest on the passing the L1 range of navigation satellites signals

Anatoly Sorokin[1],*Alexander* Borisov[2],*Mikhail* Reushev[1],*Victor* Ivanov[2],*Dmitriy* Kharlamov[1]

[1]Federal Research Center KSC SB RAS, Krasnoyarsk, Russian Federation

[2]Institute of Forest named after V.N. Sukacheva FRC KSC SB RAS, Krasnoyarsk, Russian Federation

Abstract. Experimental data on the passing of signals from navigation satellites of the L1 range through forest plantations with an anisotropic horizontal tree arrangement structure are presented. A feature of the studied environment is an irregular distance between trees in rows and a constant distance between rows. Signals were recorded by an antenna located inside the forest at heights of 0.5 and 10 meters. Based on the results of processing the amplitude-time dependences of the recorded signal by means of fast Fourier transform, qualitative differences were revealed depending on the orientation of the probe signal path in the forest stand.

1 Introduction

The formation of forest stands in forestry practice is carried out by planting seedlings and is an important tool for growing forest of highly productivity. Seedlings are usually planted in rows, but the arrangement of growing trees in one row becomes irregular over time. Forest care measures can weaken competition, increase the available resource for trees of higher quality and shorten the growing time. Such activities are carried out taking into account the intended purpose of forests [1 – 3].

The competition between all the trees of the stand even-aged stands of natural origin increases almost simultaneously by the time the crowns overlap. This leads to a massive fall of trees and even the surviving trees are in a significantly weakened state. In a planted stands, due to rows structure process of competition is separated in time and occurs in two stages. At the first stage, crowns are closing in the rows and competition for lighting is intensifying. This leads to the differentiation of trees by the diameter and size of the crowns and some trees falling off. The second stage occurs at the age of closing the crowns between the rows. Therefore, the process of tree fall in natural forests and forest plantations proceeds in different ways as a result of the difference in the horizontal structure of stands. It is necessary to control the density of the stand and the amount of wood stock, regardless of the horizontal structure, and timely cut down weakened and stunted trees. The particular importance in the cultivation of forests is getting operative information about the horizontal structure of the stand and the assessment of the ratio of biomass fractions of crowns and tree trunks. These data are available using instrumental methods for estimation the attenuation of satellite navigation signals during passing through the forest. The characteristics of the trunk wood and tree crowns affect the parameters of the NS signals during the passage of signals through the forest canopy with an azimuth along or across the rows.

This paper presents the results of monitoring the anisotropy of the horizontal structure of pine forest plantations at the age of 72 years using signals from navigation satellites (NS) of the L1 range.

2 Object of study

The experimental test site is the pine plantations of the 1st quality class, located on the territory of urban forests near Akademgorodok. The plantations are 72 years old. The size of the plot is 480x90 meters. The main taxation characteristics of the pine stand are as follows; stand density 1425 pcs / ha; average tree height (hav) - 25 m; the average diameter at breast height - 24.2 cm; stock of stem wood - 743 m^3 / ha; the sum of the cross-sections of the trunks - 65.5 m^2 / ha; stand fullness -1.47. Pines were planted in rows oriented along an azimuth of 72° with a row spacing of 3 meters. The distance between trees in rows varies from 1 to 3 meters with an average distance between trees in a row of 1.5 m. The stand density in the test plot significantly exceeds this parameter for a normal stand at this age - 757 pcs / ha [4]. Competition arises between trees for available resources, which leads to a weakening of the stand as a whole and a smaller average diameter of trunks with poorly developed crowns. The bulk of trees (more than 80%) in the classification according to the degree of domination and oppression belongs to the III and IV classes of Kraft [5].
The top view of the site is shown in Figure 1a. The internal structure of the stand with an azimuth along or across the rows is shown in Figures 1b and 1c.

Fig 1. View of the fragment of the test site from above - a), ground image of the stand, along the rows - b), perpendicular to the rows - c).

The horizontal structure of the stand has a pronounced anisotropy. The spacing between rows is twice the average spacing between trees in a row. In the process of growth, a layered structure of the distribution of trees in the form of rows appears.

In the vertical direction, the stand also has a layered structure. An estimate of the ratio of the effective average density of biomass in the layers of pine crowns and trunks gives a value of 1/16 [6]. The vertical complete orientation ordering of the trunks and partial branches of the crowns is realized in the stand. There are significant differences in the electrophysical properties of crown layers and tree trunks - biomass density and anisotropy of the effective dielectric constant of the stand.

3 Passing of NS signals through the pine stand

The attenuation of NS signals passing through a forest area depends on the size of the trunks and the density of the trees- stand. The stand fullness of the test site exceeds by 47% that for a "normal trees-stand" [4].

In the vertical direction, the tree stand has a layered structure. The top layer, formed by tree crowns, is at a height of 15 to 20 meters. The average crown height is 5.4 m, the trunk layer 19.6 m. Estimation of the ratio of the average biomass density of the crown and trunk layers gives a value of 1/16 [4]. Vertical orientational ordering of trunks and partial ordering of crown branches are realized in the tree stand. There are differences in the electrical properties of the crowns and trunks of the trees-stand: the density of biomass, the vertical extent of the magnitude and the anisotropy of the dielectric constant.

Measurements of the characteristics of signals from the NS systems GLONASS, GPS, Galileo in the process of propagation in the trees-stand were carried out using a specialized four-channel receiver. The receiving antenna recording navigation signals with right-hand circular polarization was placed at heights of h_1 = 10 m and h_2 = 0.5 m from the earth's surface. The trajectories and the length of the paths of the NS signals in the stand were determined from the data of the location of satellites, the length of the layers of crowns and trunks, and the heights of the antenna above the soil.

The measurement scheme is shown in Fig. 2.

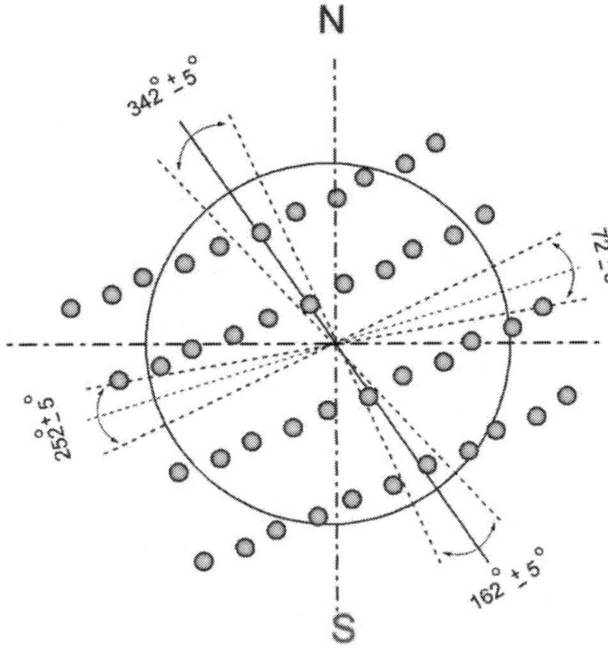

Fig 2. Scheme of measurements of navigation signals on the test site.

The antenna pattern was oriented vertically. The receiver recorded the values of the magnitude proportional to the power of the transmitted signal, averaged over 1 second in the form of amplitude-time dependences (ATD). The set of signals from the NS recorded by the receiver made it possible to determine the coordinates of the sounding volume of the stand and the coefficients of linear attenuation with a coordinate reference [7].

4 Results and discussion

Examples of measured individual ATD at antenna heights h_1 and h_2 are shown in Figures 3 and 4 in the form of normalized (per unit) functions S(t).

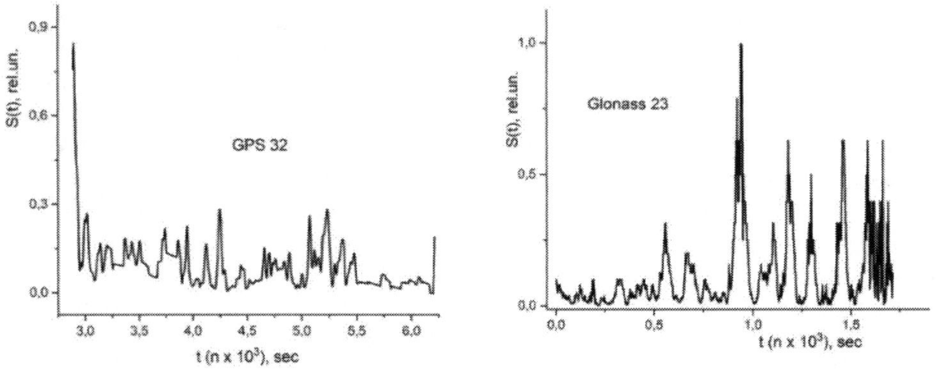

a. b.

Fig 3.ATD, antenna height h_1 = 10 m:(a.) NS GPS - 32 (azimuth ≈ 342°, elevation ≈30°); (b.) NS GLONASS - 23 (azimuth ≈ 250°, elevation ≈ 30°).

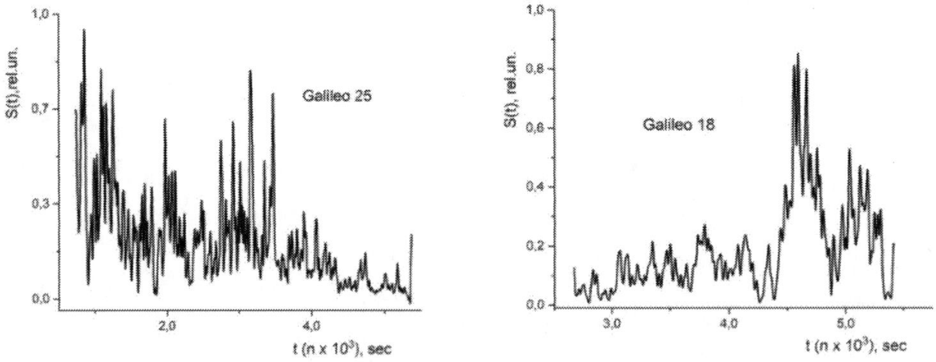

a. b.

Fig 4. ATD, antenna height h_2 = 0.5 m:(a.) NS Galileo– 25 (azimuth ≈ 160°, elevation ≈30°);(b.) NS Galileo – 18 (azimuth ≈ 260°, elevation ≈ 30°).

As can be seen from Figures 3 and 4, the form of the S (t) dependences for the NS at a height close to the ground surface (h_2 = 0.5 m) differs from the form of the S(t) dependences obtained during measurements at a height near the lower boundary of the tree crown layer (h_1 = 10 m) for directions of signal propagation along (=) and across (+) tree rows.

Analysis of S(t) with the use of fast Fourier transform (FFT) determined qualitative regularities in the registration of NS-signals in the orthogonal directions (=) and (+) of the trees-stand structure. Figures 5 and 6 show the results of processing S (t) in the form of dependences S (ν).

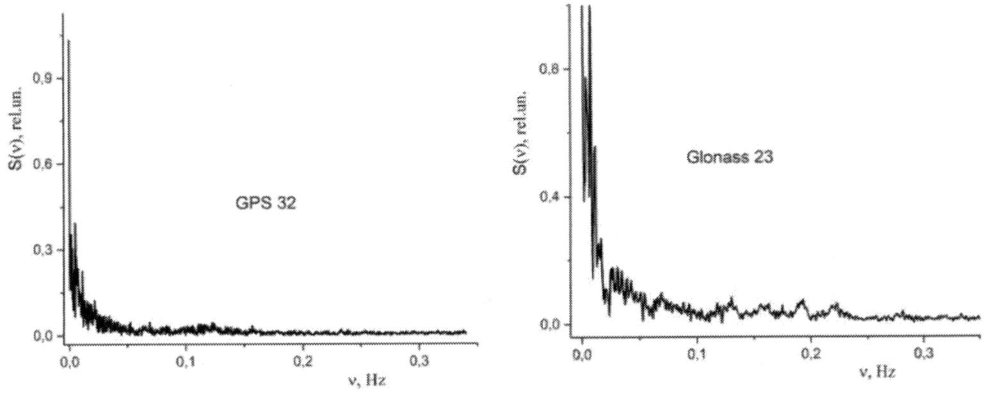

a. b.

Fig 5. FFT, antenna height h_1 = 10 m:(a.) NS GPS - 32 (azimuth ≈ 342°, elevation ≈30°); (b.) NS GLONASS - 23 (azimuth ≈ 250°, elevation ≈ 30°).

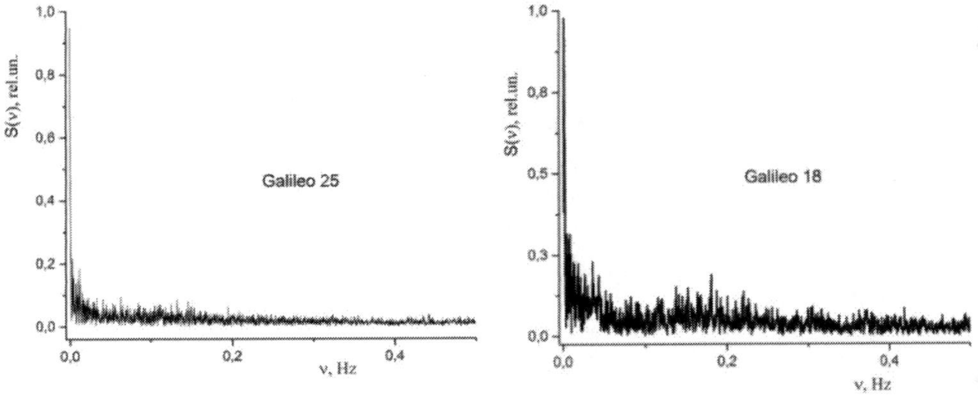

a. b.

Fig 6. FFT, antenna height h_2= 0.5 m:(a.) NS Galileo – 25 (azimuth ≈ 160°, elevation ≈30°);(b.) NS Galileo – 18 (azimuth ≈ 260°, elevation ≈ 30°).

It can be seen from Figures 5 and 6 that the Fourier spectra of NS have differences when they propagate in a forest along and across rows of trees. Similar differences were observed earlier in measurements using GNSS – reflectometry of various surfaces of the earth's covers [8]. Subsequent analysis made it possible to interpret the results obtained in the form of a function S (ν):

$$S(\nu) = S_0 \nu^{-\alpha} \quad (1)$$

Here S_0 is the maximum amplitude of the Fourier harmonics, the amplitude of the signal S(ν), α is the exponent depending on the structure of the trees- stand (crown, trunks) of the studied taxation area, ν – is the frequency in Hz.

In the examples under consideration (Figures 5 and 6), the coefficients of the functional dependence S(ν) are given in Table 1.

Table 1. Coefficients of dependence S (ν).

№	Наименование НС	S_0	α	*h*, м	НСД
1	GPS – 32	0.78	0.77	10	(+)

2	GLONASS – 23	0.50	0.81	10	(=)
3	Galileo – 25	1.86	0.47	0.5	(+)
4	Galileo – 18	71.09	0.43	0.5	(=)

As can be seen from Table 1, the attenuation coefficients of the function S(ν) when registering signals from the NS in the crown layer, (h_1 = 10 m) in relation to the signals recorded in the trunk layer (h_2 = 0.5 m).

Similar differences in the coefficients α were also observed when the NS was recorded by the receivers at heights of h_1 = 10 m and h_2= 0.5 m from other satellites. The results of the research allow us to conclude that the observed qualitative differences in ATD followed by processing using a fast Fourier transform can determine the parameters for a quantitative assessment of the biometric characteristics of the trees-stand.

Conclusions

Measurements of the amplitude-time dependences of the signals of navigation satellites in the layers of crowns and trunks of pine plantations with an anisotropic horizontal structure of tree arrangement in the form of parallel rows have been carried out.

In the array of experimental data, a sample of trajectories of the direction of signals passing along and across the rows of trees was carried out. The differences in the amplitude-time dependences of such signals are revealed.

The processing of the amplitude-time dependences of the signals using the fast Fourier transform has been carried out. The differences in the Fourier spectra were revealed depending on the orientation of the path of the sounding signal in the forest stand.

A quantitative interpretation of the Fourier spectra in the form of an empirical exponential function is proposed.

References

1. Borisov, A.N., Ivanov, V.V. and Petrenko, A.E., 2019. Formation of the spatial structure of pine stands during thinning. Forestry, (1), pp. 7-18.
2. Nagimov, Z.Ya., 2000. Regularities of growth and formation of aboveground phytomass of pine stands (Doctoral dissertation, Ural State Forestry University).
3. Rogozin, M.V. and Razin, GS, 2015. Development of forest stands. Models, laws, hypotheses.
4. Shvidenko, A.Z., Shchepashchenko, D.G., Nilsson, S. and Bului, Yu.I., 2008. Tables and models of growth and productivity of planting of the main ec-forming rocks of northern Eurasia.
5. Eitingen G.R. Forestry. Moscow .: Selkhozgiz. 1949.246 s.
6. Stakanov, VD, 1990. Distribution of organic matter in different parts of Scots pine trees. Forestry, (4), p. 25.
7. Pat. 2 682 718 of the Russian Federation. IPC G01S 19/03 (2010.01) Method for determining the linear attenuation coefficients of navigation spacecraft signals in a woodland with coordinate reference // Makarov D.S., Savin I.V., Sorokin A.V., Fomin S.V. Kharlamov D.V. No. 2018119448; declared 05/25/2018; publ. 03/21/2019, Bul. No. 9. 8 p.
8. Kuzmin, E.V. and Sorokin, AV, 2021. Fourier spectra of amplitude variations of the superposition of signals from navigation satellites near surfaces with different properties. Letters to the Journal of Technical Physics, 47 (18), pp. 26-28.

Use of 3D models for control of the main pipeline protective facilities

Maria Shlyakhova[1]

[1]Siberian State University of Geosystems and Technologies, Photogrammetry and Remote Sensing Department, 10 Plakhotnogo street, 630108, Novosibirsk, Russia

Abstract. The primary purpose of main pipelines is to transport oil products from a point to the point of destination. Therefore, its length can be thousands of kilometers, most often it is located in difficult natural and climatic conditions. Main pipeline continuous operation and protection from accidents or breakdowns requires passive or active control of protective structures using 3D technologies. The article briefly describes the technology for creating 3D models based on materials from aerial laser scanning and aerosurvey.

1 Introduction

Oil products are generally transported by pipelines associated with the risk of accidents and, in turn, an oil spill. The use of 3D models to control the main pipeline protective structures is an urgent task, for it allows obtaining realistic data on the state of the surveyed object and subsequent decision making on identifying any deviations from the specified parameters. The accuracy and detail of 3D models will mainly depend on the selected information collection technology. There is a classification of protective structures depending on the type of material.

Depending on the material of a main pipeline protective structure, two main types are distinguished: earthen and concrete. The authors [1] identified the main criteria for the accuracy control (Tab. 1).

Table 1. The choice of data collection technology depending on the type of protective structure.

Type of protective structure	Accuracy in plane view	Accuracy in elevation	Data collection technology for creating 3D models of protective structures
Earthen structure	15 cm	15 cm	Aerosurvey, aerial laser photography
Concrete structure	10 mm	15 mm	Aerial and ground laser scanning

In the northern regions, the main protective structures are earthworks, which can be reasonably controlled by methods of aerial laser scanning and aerosurvey. When using this technology for collecting information, the following types of data are obtained: a digital model of the relief and terrain, as well as information about the surface texture [2, 3].

The following technological processes are used to build a 3D model (Fig. 1) [4].

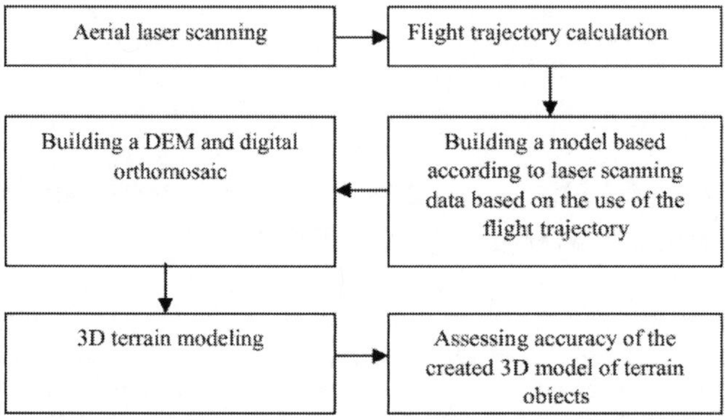

Fig. 1. Laser scanning data processing and digital aerial photography scheme for creating 3D models

2 Experiment

Stereophotogrammetric and laser surveys were performed using a Leica ALS-60 laser scanner and a Leica RCD 105 digital aerial camera to create 3D models of main pipelines and protective structures (Fig. 1). The surveys were accompanied by measurements of onboard GPS and IMU. Terra Solid software product based on the Bentley MicroStation platform was used to process the data (Figure 1).

The images were used to measure the plane coordinates of the object image points and to calculate spatial coordinates of the object points (X, Y, Z), which are the points of the spatial terrain model. An image element belongs to each point of the object. Brightness of an element and an image for a terrain point (a pixel in a terrain) can be taken from any image belonging to a set of overlapping images in the corresponding point area. Then the coordinates of the points were recalculated into the screen plane coordinate system x, y, z, and brightness of the element with x, y coordinates was assigned previously defined brightness of the point on the terrain model with X, Y, Z coordinates.

Figure 2 shows the airborne laser scanning data processed in the Terra Solid software.

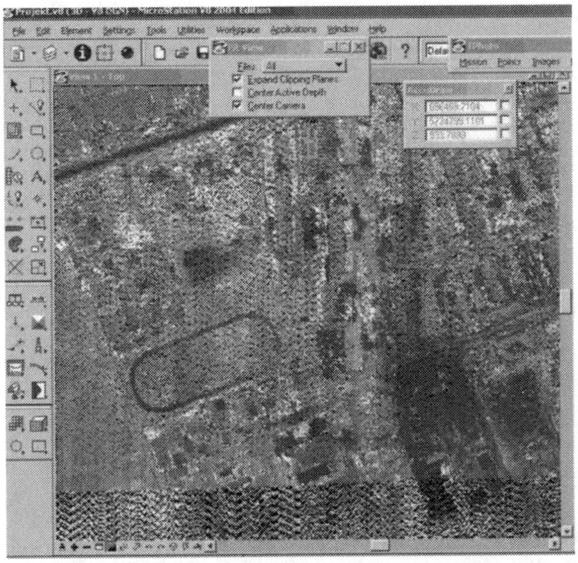

Fig. 2 - Airborne laser scanning processing data in PP Terra Solid

The next stage was building a digital elevation model and digital orthomosaic. As a result, a 3D hedgehog was created, and the accuracy was assessed by determining the points on the photographic image and the finished model, after which the affine transformation coefficients and the texture element coordinates were found. The result of evaluating the root-mean-square error accuracy is 0.33 m in elevation and 0.57 m in plane view. [5]

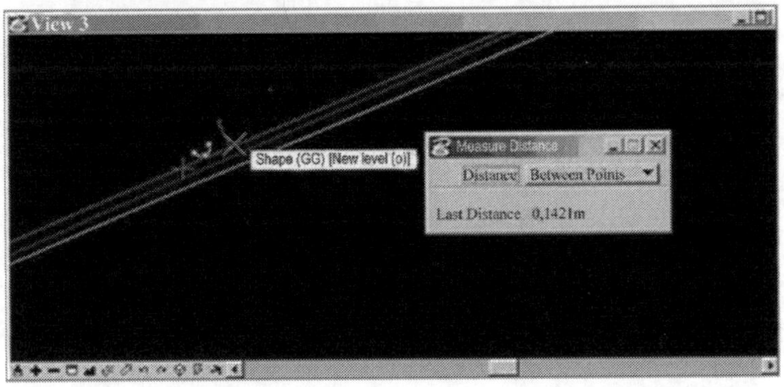

Fig. 3. Measuring the discrepancy between models.

To build a realistic 3D measurement model, it is necessary to perform a large number of measurements of characteristic points, which must also be identified in the images from which the image texture is taken. In addition to coordinate transformations, a real texture of the image should be formed when building 3D models.

In most modern technologies, the texture is oriented relative to the object surface by identifying image points on the texture and in a 3D mode. Then affine transformation coefficients are determined and coordinates of the texture elements are recalculated into the corresponding elements of the 3D model. Thus, the texture only approximately corresponds to the object and is not a measurement one. In addition, standard aerial survey (with 40% and 60% overlap) is not sufficient for texture generation, so additional images are required to create a realistic model.

Thus, knowing the accuracy of the protective structure control and the choice of survey methods, we can create 3D models with maximum accuracy in order to timely identify potentially dangerous sections of the pipelines and reduce the risk of accidents.

The information obtained can be used to create 3D models that will help to timely identify emergencies, and control reliability of the pipeline by monitoring it throughout the entire period of operation. 3D models also make it possible to track changes and automatically simulate situations in the three-dimensional form of the deformation area.

Thus, the use of 3D models helps:

- Make automatic measurements of geometric parameters of communication;
- Exercise control over the raw material transportation and extraction;
- Update information based on terrain change data;
- Monitor the state of communications;
- Simulate emergency situations based on the various sensors readings (pressure, temperature) and prevent them.

It should be noted that raw materials can be transported in various ways, but the pipeline is still the most profitable and safe way. In order to avoid emergency situations, it is necessary to carry out 3D monitoring using remote sensing methods, and to build 3D models with maximum accuracy and detail based on the information obtained. With the help of such models, problems are identified and analyzed at pipeline sections in order to quickly eliminate accidents or breakdowns. Therefore, environmental disasters can be avoided.

The work was carried out within the framework of the state assignment of the Ministry of Education and Science of Russia (subject: Development of theory and technological solutions for monitoring the state of protective structures when pumping oil products using active remote sensing methods, No. 0807-2020-0002).

References

1. A.V. Komissarov, M.M. Shlyakhova, M.A. Altyntsev, E.N. Kulik. Criteria for the main pipeline protective structure control. *SSUGT Bulletin*. 2020. V. 25, No. 4, pp. 96–104.
2. V.V. Dedkova, M.M. Shlyakhova. Monitoring of the main pipeline technical condition of using remote sensing methods. *Regional problems of the Earth remote sensing of: proceedings of the VII International Scientific Conference*, Krasnoyarsk, 29 Sept. – 2 Oct. 2020, pp. 192–195.
3. M.M. Shlyakhova, V.V. Dedkova. Prospects for the use of aerial surveys for the control of main pipeline protective structures. *Regional problems of the Earth remote sensing of: proceedings of the VII International Scientific Conference*, Krasnoyarsk, 29 Sept. – 2 Oct. 2020, pp. 316–319.
4. A.P. Guk, M.M. Shlyakhova. Development of methods for creating 3D models from aerospace images of high and ultra-high resolution and other remote sensing data. *Izvestiya VUZov. Geodesy and aerial photography*. 2011. No. 2, pp. 32-34.
5. A.P. Guk, M.M. Shlyakhova. Some problems of building realistic 3D measurement models from remote sensing data. *SSUGT Bulletin*. 2015. V. 2, No. 4, pp. 51-60.
6. A.P. Guk, M.M. Shlyakhova, R.V. Brezhnev. Potential of using realistic 3D models built from digital images to solve a wide range of problems. *National scientific and*

practical conference Regulation of Land and Property Relations in Russia, 2018. SSUGT. V. 1, pp. 172-176.

Spatial-time variations in the distribution of atmospheric methane at high latitudes of the northern hemisphere

Sardana Vasileva

[1]Yu.G. Shafer Institute of Cosmophysical Research and Aeronomy of Siberian Branch of the Russian Academy of Sciences, Yakutsk, Russia

Abstract. A preliminary study of spatio-temporal variations of methane distribution at high latitudes of the northern hemisphere was conducted based on AIRS and JR-STATION data. The analysis showed a noticeable increase (~4%) in the methane concentration in the lower layers of the atmosphere and upper layers of the troposphere. According to the AIRS data, winter (February) and summer (July- August) peaks of the methane concentration were reported in the seasonal cycle of 2004-2018. Seasonal methane variations in the upper troposphere obtained from AIRS data have similarities to ground observations, but with a smaller amplitude

Introduction

In recent decades, significant warming has been observed at high latitudes of the northern hemisphere [1]. Climate change has a significant impact on methane (CH4) sources. The global warming potential of CH4 exceeds 20 times the warming potential of CO2 [2]. The air temperature directly affects the sources of CH4 in the soil by affecting the metabolic rate of methanogens [3], as well as indirectly - through the melting of permafrost and changes of groundwater's depth and snow cover characteristics [4]. Other climatic changes in the Arctic may increase the intake of CH4 into the atmosphere due to the possible destabilization of methane hydrates in oceanic sediments [5].

The rapid pace of climate change in high latitudes and the expansion of anthropogenic activity in the Arctic in the future may lead to a significant increase in CH4 emissions into the atmosphere. Which can contribute to an increase in the rate of heating of the atmosphere. Under the current conditions, it is very important to conduct studies of variations in atmospheric methane.

The aim of this work is to study the spatiotemporal variations in the distribution of atmospheric methane at high latitudes of the northern hemisphere.

Materials

To analyze spatio-temporal dynamics of CH4 concentration during 2004-2018 data from the AIRS (Atmospheric Infrared Sounder) radiometer installed on the AQUA satellite (AIRX3STM product, https://giovanni.gsfc.nasa.gov) was used. The data for the 400 hPa pressure level were used in this study, due to the sensor's maximum sensitivity to CH4 at this

https://doi.org/10.1051/e3sconf/202133302001

level. In the middle and high latitudes of the Northern Hemisphere, it reaches 400-300 hPa (about 7-9 km) [6].

Data from the JR-STATION greenhouse gas monitoring network for 2004-2018 were also used in the analysis (http://db.cger.nies.go.jp/portal/). The greenhouse gas-monitoring network is located in Western Siberia. [7].

Results and discussion

An area covering a part of the West Siberian Plain and the Central Siberian Plateau (60°-73° s.w., 68°-126° v.d.) was selected for research. In Fig. 1, black lines limit the sector of the studied area, the locations of four ground stations of the JR-STATION network are shown by dots: Igrim, Noyabrsk, Demyansk, Karasevoye.

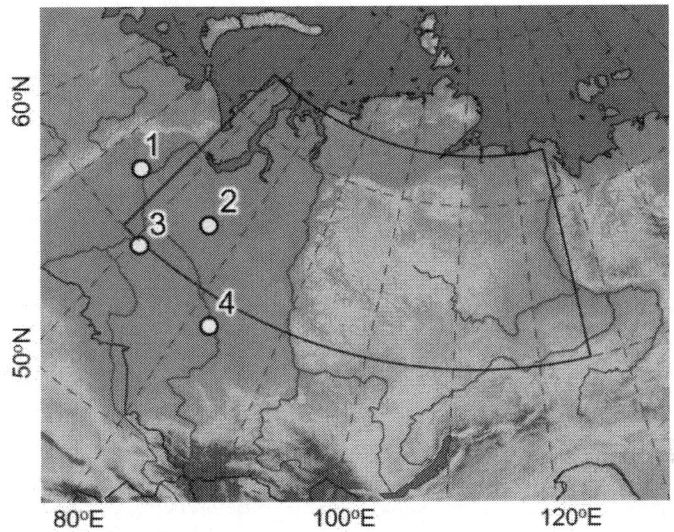

Fig. 1. Location of the test site and ground stations: 1- Igrim, 2- Noyabrsk, 3-Demyansk, 4-Karasevoye

Inter-annual and seasonal variations of methane concentration from satellite and ground data are presented in Fig. 2a. As it can be seen from Fig. 2a, the concentration of CH4 increases at the surface level and upper troposphere; the total increase during the study period according to AIRS data was ~ 67 ppb, according to ground data ~ 86 ppb.

Fig. 2b shows seasonal variations in CH4 content from satellite and ground data for 2004-2018. There are two maxima in the surface layer: winter (February) and summer (July-August) and two minima: spring (May) and autumn (October) [7, 8]. The methane concentration increase in the summer period has been recorded since the beginning of May, when the soil temperature increases and contributes to the increase in the activity of methanogenic microorganisms and reaches its peak in July-August, showing the seasonal CH4 emissions from swamp ecosystems [9]. Seasonal variations of methane content in the upper troposphere are similar to variations in the lower troposphere but have smaller amplitude. In the upper troposphere, there is also a winter maximum (January) and a summer increase in methane concentration. The spring minimum of concentration also occurs in May, but the winter minimum (December) is less pronounced. One of the explanations for the summer increase of methane in the upper troposphere may be an increase in vertical transport from the lower troposphere in summer [10]. Also in [11], the authors showed that in the summer, remote sources in Europe, North America, and Asia may contribute to the methane content.

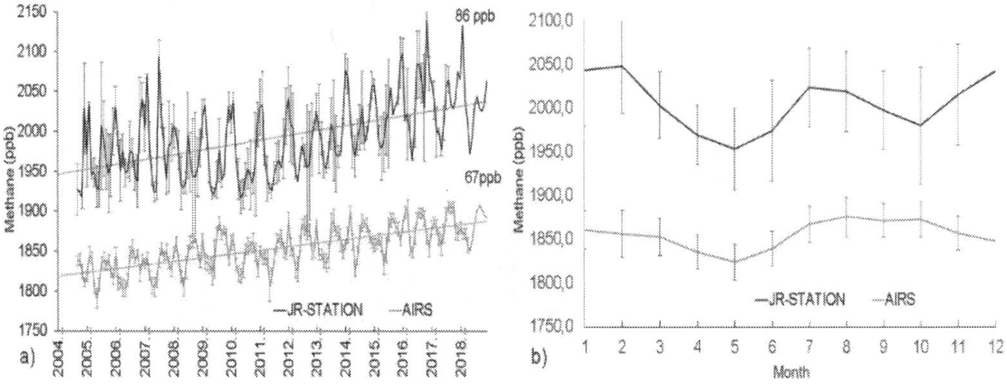

Fig. 2: a) Inter-annual variations of methane concentration, b) seasonal course of methane concentration according to AIRS data at 400 hPa pressure level and averaged for 4 stations JR-STATION data

Conclusion.

Analysis of spatio-temporal variations in the distribution of atmospheric methane at high latitudes of the Northern Hemisphere in 2004-2018 showed that there is a noticeable increase (~4%) in the concentration of methane in the lower atmosphere and upper troposphere. According to the ground data, the seasonal cycle of methane concentration during the study period has winter (February) and summer (July- August) peaks. In the upper troposphere, seasonal variations in methane content have similarities, but also a smaller amplitude. In the upper troposphere, there is also a winter maximum (January) and a summer increase in the concentration of methane. The spring minimum of concentration also occurs in May, but the winter minimum (December) is less pronounced.

References

1. Climate Change 2013: The Physical Science Basis. IPCC Working Group I Contribution to AR5. P. 465–570 (Bern, Switzerland, 2016).

2. Отчет Межправительственной группы Экспертов по Изменениям Климата (МГЭИК), 2007 // http://www.ipcc.ch/pdf/assessment-report/ar4/syr/ar4_syr_ru.pdf.

3. Christensen, T. R., Ekberg, A., Ström, L., Mastepanov, M., Panikov, N., Öquist, M., Svensson, B. H., Nykänen, H., Martikainen, P. J., and Oskarsson, H.: Factors controlling large scale variations in methane emissions from wetlands, Geophys. Res. Lett., 30, 1414, doi:10.1029/2002GL016848, 2003.

4. Khvorostyanov, D.V., Ciais, P., Krinner, G. and Zimov, S.A. Vulnerability of east Siberia's frozen carbon stores to future warming. Geophys. Res. Lett. 35, doi: 10.1029/2008GL033639. 2008.

5. Shakhova, N., Semiletov, I., Salyuk, A., Yusupov, V., Kosmach, D., and Gustafsson: Extensive Methane Venting to the Atmosphere from Sediments of the East Siberian Arctic Shelf, Science, 327, 1246–1250, doi:10.1126/science.1182221, 2010.

6. Xiong X., Barnet C., Maddy E., Sweeney C., Liu X., Zhou L. and Goldberg M. Characterization and validation of methane products from the Atmospheric Infrared Sounder (AIRS) // Journal of Geophysical Research. Vol. 113. 2008. P. 1-14.

7. Sasakawa M., Shimoyama K., Machida T. et al. Continuous measurements of methane from a tower network over Siberia. *Tellus B*, 2010, vol. 62, pp. 403–416.

8. Arshinov, M.Y., Belan, B.D., Davydov, D.K. et al. Spatial and temporal variability of CO2 and CH4 concentrations in the surface atmospheric layer over West Siberia. Atmos Ocean Opt 22, 84–93 (2009). https://doi.org/10.1134/S1024856009010126

9. Kim, H.-S.; Maksyutov, S.; Glagolev, M.V.; Machida, T.; Patra, P.; Sudo, K.; Inoue, G. Evaluation of methane emissions from West Siberian wetlands based on inverse modeling. *Environ. Res. Lett.* 2011, *6*, 035201.

10. Xiong, X., Barnet, C. D., Zhuang, Q., Machida, T., Sweeney, C., and Patra, P. K. (2010), Mid-upper tropospheric methane in the high Northern Hemisphere: Spaceborne observations by AIRS, aircraft measurements, and model simulations, J. Geophys. Res., 115, D19309, doi:10.1029/2009JD013796.

11. Мордвин, Е. Ю. Метан в атмосфере Западной Сибири : [монография] / Е. Ю. Мордвин, А. А. Лагутин ; АлтГУ. - Барнаул : [Азбука], 2016. - 148 с. : ил.

Influence of temperature inversions on air pollution in Krasnoyarsk

Alexander Dergunov[1*], *Oleg* Yakubailic[2]

[1]Federal Research Center Krasnoyarsk Science Center SB RAS, Krasnoyarsk, 660036, Russian Federation
[2]Institute of Computational Modelling SB RAS, Krasnoyarsk, 660036, Russian Federation

Abstract. The paper analyzes the meteorological conditions in Krasnoyarsk in the period from 2019 to 2020. The relationship between the occurrence of temperature inversions in the surface layer of the atmosphere and air pollution by suspended solid particles ($PM_{2.5}$) is investigated. The paper uses a set of meteorological data of the NCEP GFS weather forecast model on air temperature on three isobaric surfaces (1000, 925 and 850 mb), wind gusts and the height of the atmospheric boundary layer. The air monitoring system of the KSC SB RAS provides data on solid particle concentrations $PM_{2.5}$ and wind speed. The relationship between the presence of temperature inversions in the lower layer of the atmosphere and periods of significant air pollution is shown, as well as the dependence of changes in wind speed and the height of the boundary layer of the atmosphere with changes in the average daily $PM_{2.5}$ concentration. The possibility of using the meteorological data of the NCEP GFS model to study the surface layer of the atmosphere and the periods of its pollution, predicting the deterioration of the environmental situation in Krasnoyarsk, is demonstrated.

1 Introduction

Atmospheric pollution is one of the main problems of large cities in the world. As a result of the anthropogenic factors, a large amount of harmful substances that pollute the air of large metropolitan areas enter the atmosphere [1]. Studies show an association between increased concentrations of particulate matter (PM) in the air and the deterioration of human health [2]. During periods of unfavorable meteorological conditions (UMC), there is a sharp increase in the concentration of pollutants in the lower layer of the atmosphere. We are talking about industrial and automobile emissions, furnace heating, and so on [3]. One of the determining factors of UMC is the temperature inversion in the atmosphere. This is a layer of warm air that traps the vertical dispersion of harmful impurities [4].

There is a need to study atmospheric processes with the help of meteorological data sets to analyze periods leading to environmental degradation.

Sets of meteorological data can be obtained, for example, from ground-based environmental monitoring stations. The advantages of such weather information include a

* Corresponding author: alexdergunov@icm.krasn.ru

fairly high accuracy of measurements on the spot. However, the disadvantages of such data are that basic parameters limit such data sets, and the ground stations themselves can be located at a great distance from each other, which imposes certain restrictions on the spatial coverage of the problem being solved.

An alternative option is the weather data of various weather forecast models, for example, the Global Forecast System model. These sources provide many different layers of meteorological information on dozens of vertical levels, plotted on regular rectangular grids covering the entire Globe. The spatial resolution of such data can vary from 2.5 to 0.25 degrees and higher [5]. Developers try to improve this parameter regularly.

The purpose of this work is to analyze the influence of temperature inversions in the surface layer of the atmosphere on the formation of an unfavorable environmental situation in the city of Krasnoyarsk in 2019-2020 based on high-spatial-resolution meteorological data of the NCEP Global Forecast System weather forecast model. Identification of the relationship between the periods with the average daily concentrations of $PM_{2.5}$ particulate matter suspended in the air, exceeding the average daily maximum permissible concentrations (MPC), the speed, wind gusts, and the height of the atmosphere's boundary layer.

The research area in this work is the city of Krasnoyarsk and part of the adjacent territories (Fig. 1).

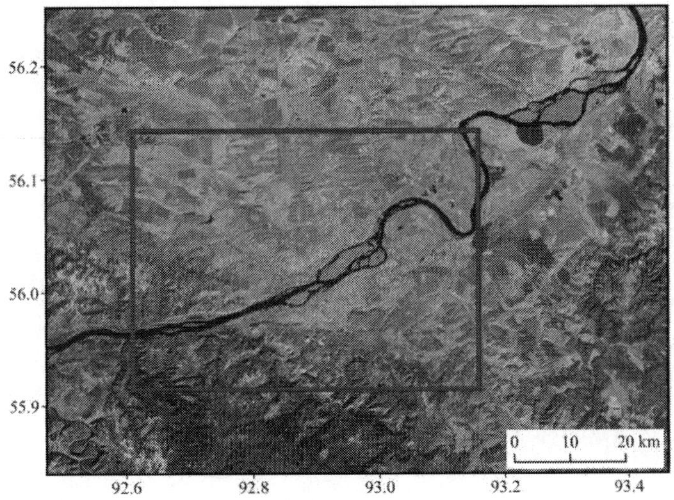

Fig. 1. The study area is the city of Krasnoyarsk.

12 periods of UMC were established in the city of Krasnoyarsk from 2019 to 2020 according to the Ministry of Ecology and Rational Nature Management data of the Krasnoyarsk Territory [6]. The periods of UMC are characterized by low wind speed and high particulate matter concentrations, significantly exceeding the average daily MPC equal to 0.035 mg/m^3.

2 Materials and methods

The work uses meteorological information from the dataset of the Global Forecast System (GFS) model. This is a numerical weather forecasting system containing a global computer model and variational analysis performed by the US National Weather Service. This model combines four separate models: atmosphere, ocean, land/soil, and sea ice. Dozens of atmospheric and ground-soil variables are available in this data set, from temperature, wind, and precipitation, to soil moisture indicators and atmospheric ozone concentration. This is one of the most famous meteorological models in the world. Global data analysis and forecasting are carried out 4 times a day. The weather forecast is available up to 16 days in advance [5].

The accuracy of the GFS model is constantly improving. In particular, data with a horizontal resolution of 1 degree has been available since March 2004, with a resolution of 0.5 degrees since January 2007. Since January 2015, the horizontal resolution has been 0.25 degrees (about 25 km at the latitude of the city of Krasnoyarsk).

The GFS model data is presented in the grib2 format. This data format is standardized by the World Meteorological Organization (WMO) and is intended for storing historical and forecast weather data [7]. Each file contains more than 500 layers of various meteorological information on more than a hundred vertical levels.

This work used the actual analysis data on the air temperature at three vertical levels corresponding to three isobaric surfaces: 1000, 925, and 850 mb. Data on the height of the atmospheric boundary layer and information on wind gusts from January 1, 2019, to December 31, 2020, obtained from the GFS model data set, were also analyzed.

Inversion layers were detected by obtaining the difference between temperature data at Different vertical levels when the difference was negative. For example, with a negative difference between temperatures on isobaric surfaces of 1000 and 925 mb, the inversion was considered ground-level or elevated; if between 925 and 850 mb, it was elevated or high-altitude (inversion of the free atmosphere). If the surface elevated and high-altitude inversions were recorded simultaneously, this was considered a powerful inversion.

We also used data on the concentrations of $PM_{2.5}$ in the air and wind speed according to ground-based monitoring stations obtained at the ICM SB RAS geoportal. Data on $PM_{2.5}$ concentrations were averaged per day for all available observation posts in the city. Thus, the average daily values of the $PM_{2.5}$ concentration for the entire city of Krasnoyarsk were obtained.

3 Results and discussion

Based on the analysis of the GFS data on air temperature at various vertical levels, a comparison was made of the number of days with temperature inversions and days when the average daily $PM_{2.5}$ concentration exceeded the average daily MPC, by months for 2019 and 2020 (Fig. 2).

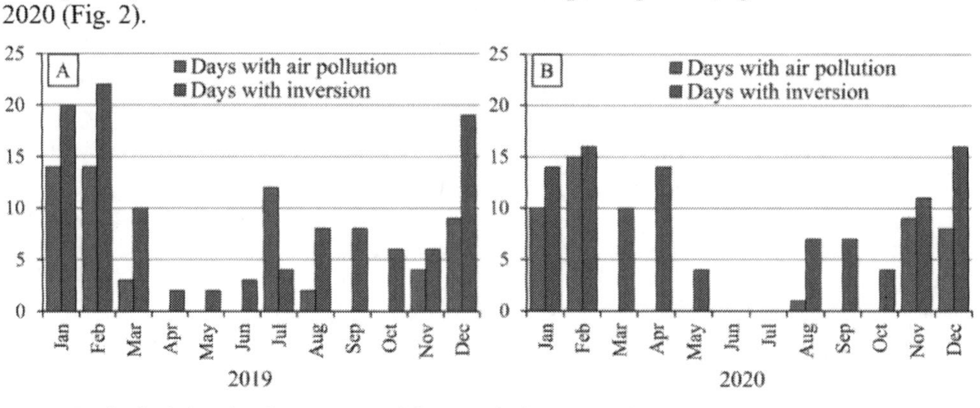

Fig. 2. Ecological situation in Krasnoyarsk by months in 2019 and 2020.

In April, May and November, the number of days with inversions is 19 days more in 2020, in the rest of the months - on the contrary, with the exception of March, when both in

2019 and in 2020 their number was 10 days. In 2019, there were 15 more days with "pollution" of the atmosphere than in 2020.

The total number of days with temperature inversions in 2019 was 110, in 2020 - 103 days. The total number of days with "pollution" of the atmosphere in 2019 was 58, in 2020 - 43 days.

Also, data analysis showed a stable relationship between changes in the average daily $PM_{2.5}$ concentration and wind gusts according to the GFS model. As an example, the period from January to February 2020 is shown (Fig. 3). This period is characterized by the fact that out of 60 days within two months, 24 days were recorded serious air pollution with $PM_{2.5}$ particulate matter.

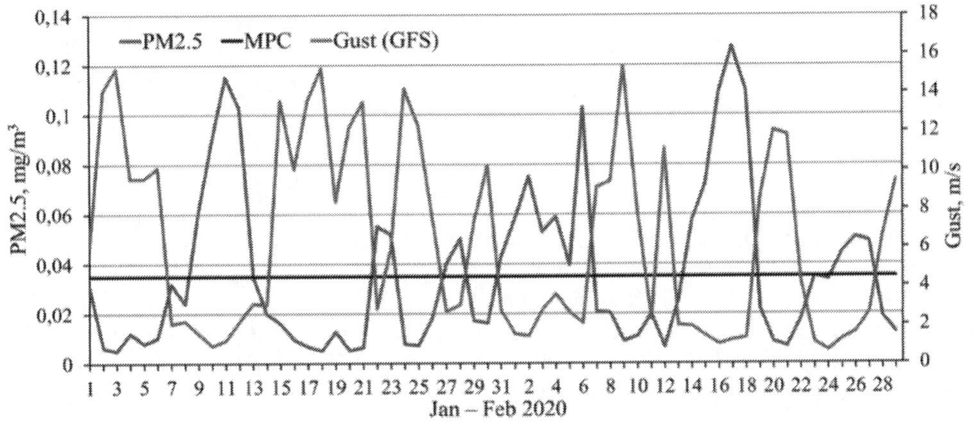

Fig. 3. Meteorological situation in Krasnoyarsk in January-February 2020. The red graph indicates the change in the average daily concentration of $PM_{2.5}$, the blue graph - the change in the speed of wind gusts. The horizontal line indicates the average daily MPC, equal to 0.035 mg/m³.

Periods of increased $PM_{2.5}$ concentration in January-February 2020: 09-13.01, 22-23.01, 27-28.01, 31.01, 01-06.02, 14-18.02 and 25-27.02. Temperature inversions were observed on these days, with the exception of January 9, February 5 and 26. It is shown that with an increase in wind speed, the level of pollution decreases and vice versa (Fig. 3).

Graphs showing the values of temperature inversions (ΔT) in the night and daytime for the period from January to February 2020 were constructed (Fig. 4).

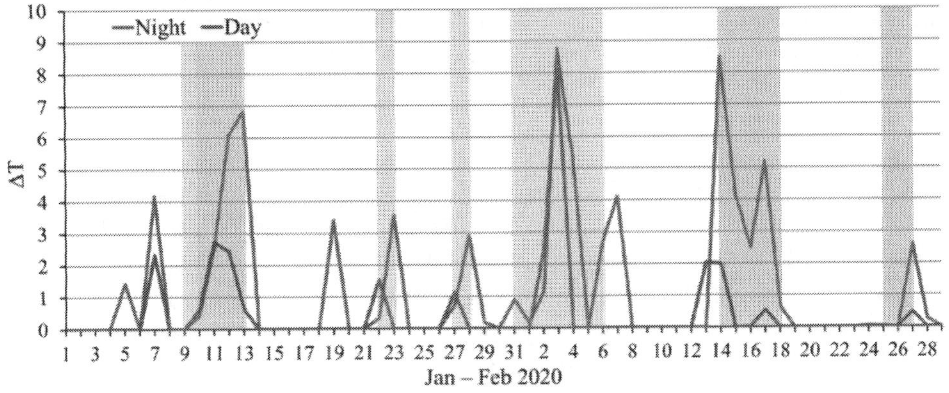

Fig. 4. Values of temperature inversions (ΔT) in January-February 2020.

In Figure 4, the blue vertical stripes mark the official periods of UMC, and the orange vertical stripes indicate the periods with an increased average daily $PM_{2.5}$ concentration identified as a result of data analysis.

Analysis of the data showed that, as a rule, the highest values of the value of temperature inversions are recorded during the marked periods of unfavorable environmental conditions in the city (Fig. 4). The prevailing type of inversion is ground-level or elevated. During the period under consideration, nighttime inversions appeared 52% more often than daytime ones.

Additionally, the variation of the height of the planetary boundary layer of the atmosphere with the wind speed obtained from the data of the ground monitoring station was compared. The correlation coefficient was 0.79. This is due to the fact that at high wind speed the stability of the atmosphere decreases, turbulence increases, which leads to an increase in the height of the boundary layer [8]. The correlation between the wind speed obtained from the ground monitoring station and the wind gusts from the GFS model is 0.82.

4 Conclusion

As a result of the work done, the relationship between temperature inversion in the surface layer of the atmosphere and the periods of significant air pollution by $PM_{2.5}$ solid particles in the city of Krasnoyarsk in 2019 and 2020 was shown.

The largest number of days with temperature inversions and increased air pollution is observed in the cold period (December, January, February). In 2019, inversions were registered in all 12 months of the year, and in 2020, there were no inversions only in June and July.

As a rule, the highest values of temperature inversions are recorded during periods of unfavorable environmental conditions in the city. The prevailing type of inversion is ground-level or elevated. At night, temperature inversions appeared 52% more often than during the day.

It is shown that with an increase in the wind speed and the height of the boundary layer, the average daily concentration of $PM_{2.5}$ decreases.

The correlation between the wind speed obtained from the ground monitoring station and the wind gusts from the GFS model is 0.82. Between the wind speed and the height of the boundary layer - 0.79. Data on the height of the boundary layer can serve as an additional indicator when studying UMC periods in Krasnoyarsk.

The analysis of meteorological data of the GFS model contributes to solving problems related to the study of the lower layer of the atmosphere, its pollution and can play an important role in more accurately identifying periods of unfavorable meteorological conditions and forecasting them.

References

1. G. Zarubin, Y. Novikov, Hygiene of the city, Medicine, Moscow (1986).
2. Y.J. Kaufman, D. Tanre, O. Boucher. Nature. **419**. №. 6903. P. 215-223. (2002).
3. A. Eremkin, I. Kvashnin, Y. Junkerov, Standardization of emissions of pollutants into the atmosphere, Publishing house of Ass. builds. universities, Moscow (2001).
4. L. Matveev, General meteorology course. Atm. Phys., Gidromet., Leningrad (1984).
5. The Global Forecast System (GFS) Documentation (2021). URL: https://is.gd/KJHKbK

https://doi.org/10.1051/e3sconf/202133302002

6. Unfavorable meteorological conditions in Krasnoyarsk. regional departmental information and analytical data system on the state of the environment of the Krasnoyarsk Territory (2021). URL: http://www.krasecology.ru/Nmu
7. NCEP WMO GRIB2 Documentation, Version 23.0.0 (2019). URL: https://www.nco.ncep.noaa.gov/pmb/docs/grib2/grib2_doc/
8. Y. Wang, H. Sayit, A. Mamtimin, J. Zhu, C. Zhou, W. Huo, F. Yang, X. Yang, J. Gao, X. Zhao. Atm. Res. 256 (2021). doi: 10.1016/j.atmosres.2021.105567

CLIMATIC ATMOSPHERIC HEAT AND MOISTURE FLUXES OF EURASIA DUE TO OCEAN CLIMATE PROCESSES

Alexei Kartushinsky[1, 2]

[1]Siberian Federal University, Krasnoyarsk, Russia
[2] Institute of Biophysics SB RAS, Krasnoyarsk, Russia

Abstract. The sea-air interactions in the troposphere give rise in changes in direction of heat and moisture fluxes due to different atmospheric pressure. The paper discusses the possibility of using the gradient method for processing satellite data to study the spatial and temporal scales of climatic and synoptic variability of dynamic processes of the atmosphere and ocean. Recent software was used for applying and analyzing satellite and contact data. The information technology has been implemented to analyze weather and climatic conditions in Eurasia using computer processing of long-term observation series of meteorological and oceanological parameters. Based on the methods of mathematical statistics, correlations in terms of surface temperature between dynamically active zones of the World Ocean have been determined. With the dynamics of the frontal zone of the Gulf Stream as an example, the problem of the influence of the structural features of the gradient field of the ocean temperature on local changes in the weather and climatic conditions of the land is discussed. The results obtained suggest that the regional features of the dynamics of the hydrological frontal Atlantic systems are derived from the geostrophic circulation in Eurasia.

1 Introduction

The ocean and the atmosphere are two self-organizing thermodynamic systems that, by their interaction, provide the climatic and weather regimes of the Earth's planetary troposphere. Short-term energy interaction occurs between the near-surface layer of the atmosphere and the surface layer of the ocean, forming large-scale (mesoscale) circulation in the upper atmosphere and lower layers of the ocean due to vertical flows of heat, moisture, salt and density stability.

The main mechanism of acceleration (or deceleration) of horizontal flows is the co-action of the force of the pressure gradient, the deflecting force of the Earth's rotation. Also, the friction force at the boundaries of the sea-air layers for transfer (or absorption) of

* Corresponding author: kartushka@mail.ru

kinetic turbulent energy is important. Jet streams in the ocean and stable hydrological fronts are thus formed, including the formation of mesoscale eddies [1].

It is well known for everyone that circulation flows in the troposphere forms areas of high (anticyclones) and low (cyclones) atmospheric pressure and atmospheric fronts with different accumulation quantity moister (precipitation) and temperature [2]. However, long-term weather forecasts are still not accurate enough due to sea temperature anomalies, especially in the surface layer of the ocean [3].

The study of the interaction of processes in the ocean and the atmosphere with the determination of the degree of climate and weather variability on terrestrial land is important feature for future scientific knowledge about Global Climate Changes.

2 Materials and methods

As the initial data, the work uses satellite data of the sea surface temperature (SST) of the ocean, measured for different periods of time by hardware with high and low spatial resolution (AVHRR MCSST, CZCS, SeaWIFS, MIRAS, AQUARIUS, MODIS, MSS: for choice 1982-2015). Also, meteorological data on observation Eurasian stations with different time periods of averaging was used. The processing and analysis of data from satellite and contact measurements were done using modern information technology and software tools. The software for the calculation of the sea surface gradient characteristics [4, 5] and a software application for analysis of weather and climatic variability were used [6].

3 Results and discussion

To determine the global climatic regime of the ocean, the cross-correlation of the temperature gradients averaged over time and space for the different area of the World Ocean identified by the dynamic activity of the frontal hydrological zones was calculated. The results are presented in Figure 1 from which features of the ocean global heat dynamic transfer, called the Conveyer Belt [7], are visible.

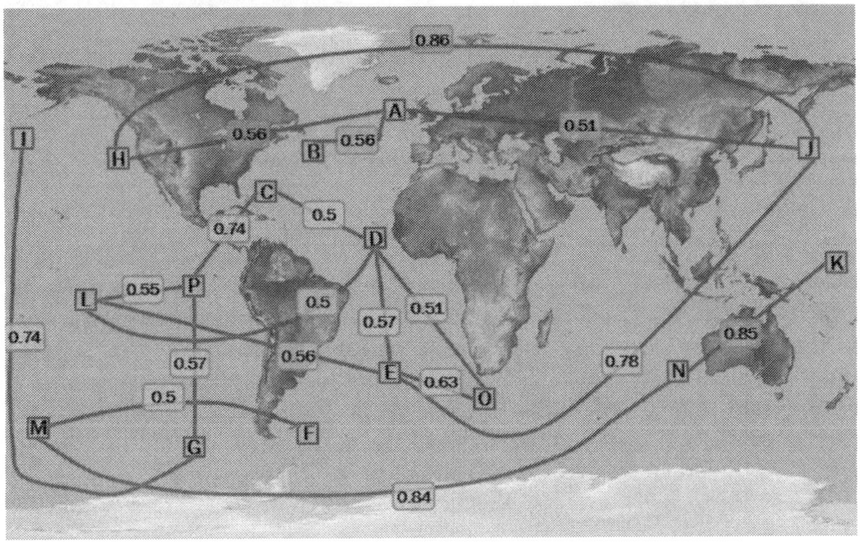

Fig. 1. Cross-correlations between sectors (dynamically active zones) on the mean average annual gradients of sea surface temperature (1982-2000).

To calculate SST gradients, we used software for the statistics of the gradient fields of the surface temperature for mean year averaging periods of time in the different ocean regions. We calculated the correlation coefficients for different zones of the World Ocean (Figure 1), which we designated by squares (letters) and the correlation coefficient (numerical).

In order to substantiate the description of the influence of regional oceanic processes formed by jet streams, the interaction zone of the Gulf Stream (warm water) and the North Labrador current (cold water) was chosen as an example (Figure 2). The structure and spatial position of this frontal zone is a trigger for regulating the flows of heat and humidity in the atmosphere and the influence of the ocean on the weather conditions of Eurasia. It is necessary to pay attention to the spatial scales of the size of frontal zone for different seasons, such as summer and winter.

Naturally, the spatial scales of the transfer of warm waters from the Gulf of Mexico to the shores of western Europe are determined by the severity of winter and the state of the ice cover. However, meandering of the Gulf Stream can reduce the flow velocity and blur the boundaries of the frontal zone, which will impede the flow of heat to the shores of Eurasia. Thus, it is necessary to monitor the structure of the gradient temperature field and its spatial scales in order to determine the influence of seasonal fluctuations in the cyclone formation regime due to evaporation in the Icelandic minimum region in the Northern Atlantic (the formation area of the low pressure atmospheric zone).

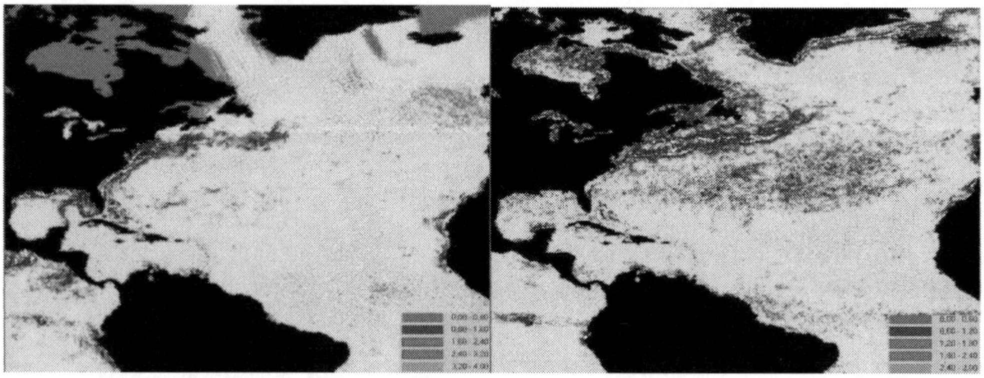

Fig. 2. Local mean seasonal temperature frontal zones of the North Atlantic (Gulf Stream zone); red – maximum value of temperature gradient. (left - winter 2000; right - summer 2000).

According to meteorological data in the structure of circulation in the North hemisphere, over the past 15-25 years, there has been a violation of stable latitudinal (zonal) transport and a transition to meridional circulation [8].

For example, the trajectories of the meridional displacement of the lower pressure area, which are commonly called "diving cyclones", moving from northwest to southeast are presented (Figure 3). For the Arctic region, such structural changes in heat and precipitation fluxes can significantly affect the climate and weather changes, which must be assessed by the speed of change of time-space scale of hydrological dynamic processes in the Northern Atlantic.

In order to show the possible influence of the trajectory of diving cyclones on climate change in the Arctic region of Russia, we used the software to create climatograms [6]. Here we use the meteorological mean monthly data of air temperature and precipitation.

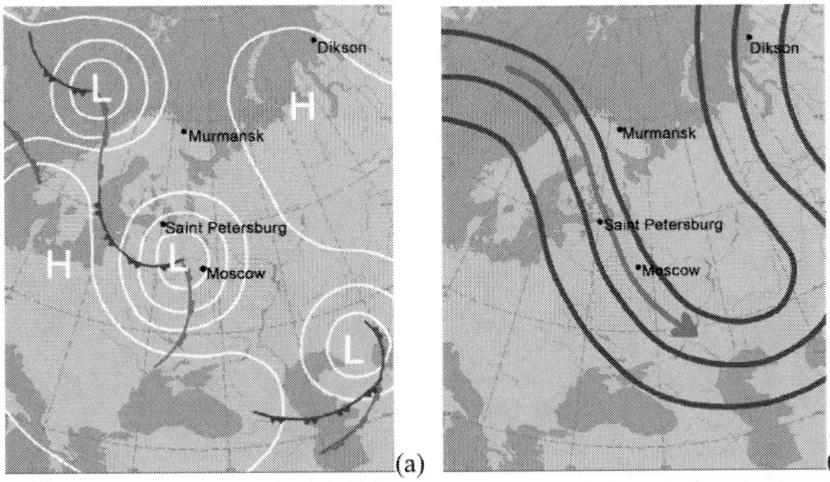

Fig. 3. The meridional trajectory of "diving cyclones" over Eurasia; a) L - area of low pressure; H - high pressure area. The warm front is shown in red, cold air is in blue; b) the main trajectory of heat and moisture flow.

For evaluation of the year oscillations in the mean monthly weather condition differences, the climatograms for different time periods of the meteorological station of the Russian Arctic region - Dikson are presented. To emphasize the significance of climate change for the period - the middle of the 20th century and the beginning of the 21st century, for comparison, we are shown the climatograms for different years (1966, 1970, 2018). (Fig. 4).

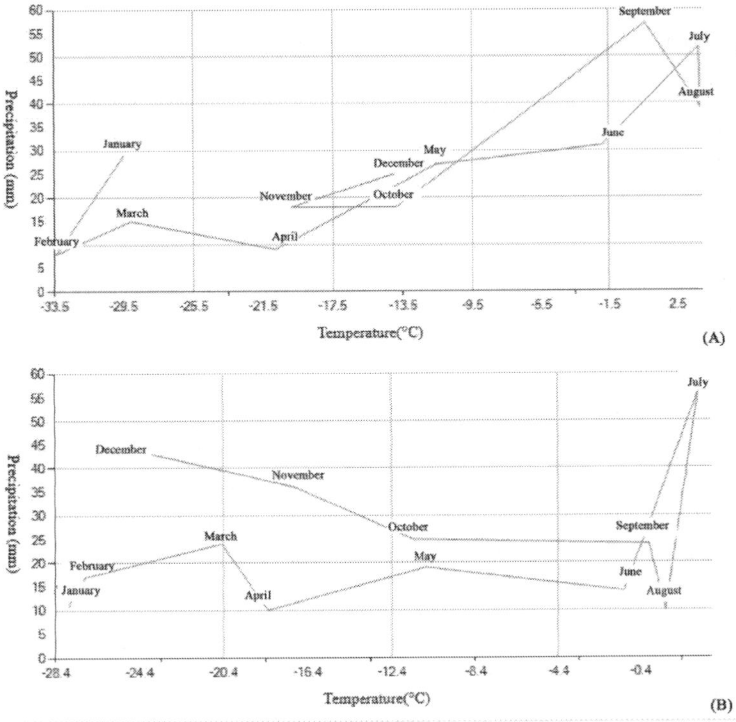

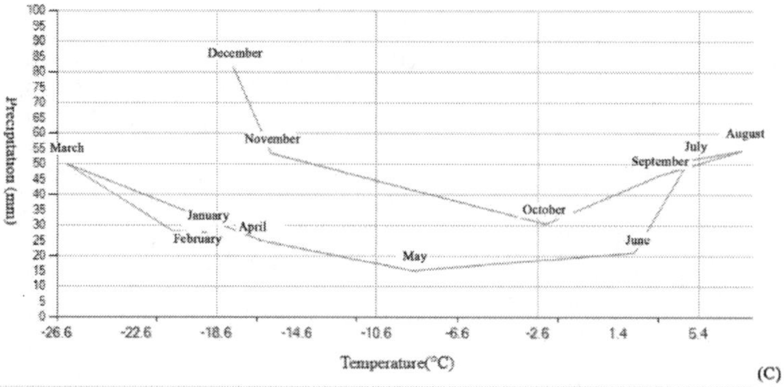

(C)

Fig. 4. Climatograms for the Dikson weather station: A - 1966, B - 1970 and C – 2018.

For the analysis of intra-annual changes in mean monthly air temperature and precipitation, climatograms with the most different shapes were selected. Climatograms for the example were selected for the interannual period of 4, 18 and 52 years for comparison with the minimum negative seasonal index of the North Atlantic Oscillations (NAO) in winter and summer (the maximum number of cyclones is 1966 and 1970) and the minimum, but positive NAO (the minimum number of cyclones in 2018) [8]. The NAO shows a direct ocean-atmosphere relationship in the North Atlantic.

For 50-55 years, there has been a shift towards an increase in the average monthly air temperature and an increase in average monthly precipitation (Figure 4C). The physical mechanism of the change in the amount of heat and moisture on the territory of Eurasia in the North Atlantic can be considered a sequence of ocean-atmosphere interaction processes: at the beginning, there is an exacerbation of the average seasonal surface temperature gradients in the frontal zone of the Gulf Stream - Labrador Current. Then there is an increase in the number of cyclones in the Icelandic minimum zone due to an increase in the amount of heat and evaporation. After that, the formed "diving cyclones" along the trajectory of the meridional direction move from the northwest to the southeast. The final stage in the action of the physical mechanism of this process is the pumping of a certain amount of heat and moisture from south to north.

4 Conclusion

Based on the results obtained, it can be assumed that the nature of the global warming lies in the background intrasecular range. The main mechanism for regulating weather conditions is the interaction of the ocean and the atmosphere at seasonal layer of variability.

The spatial position of climatic zones according to the ecological characteristic of aridity and humidity must be evaluated by the degree of interaction of the hydrological component of the ocean and the exchange of heat and moisture. The structure and spatial scales of climatic zones must be mapped based on the current state of biomes and ecosystems in terms of moisture and heat transfer regimes not on a local (regional) spatial scale, but along stable trajectories of circulation structures formed by the interaction of the ocean and the atmosphere.

The results obtained suggest that the known regional features in dynamics of the ocean-atmosphere climatic systems are derived geostrophic circulation on Eurasia. The calculations of the global ocean surface gradient characteristics based on satellite data allow us to identify areas with different dynamic activity which play a key role in determining a location of zones with a manifestation of sea-air synergism. However, one cannot ignore regional features of dynamics of ocean energy-active zones demonstration and structural features of local interaction of processes in separate areas of Global ocean and heat and moisture fluxes in the Earth troposphere.

References

1. S. Griffies, M. Winton, G.A. Whit et. al., Impacts on ocean heat from transient mesoscale eddies in a hierarchy of climate models. J. of Climate, 28, 3 (2015)
2. J. D. Hanwell, Atmospheric Processes. (1980)
3. C. E. Bulgin, C. J. Merchant, D. Ferreira, Tendencies, variability and persistence of sea surface temperature anomalies. Sci. Rep. (2020)
4. A. Kartushinsky, N. Kukoba, Investigation of gradient fields of the Earth surface based on satellite data. Vestnik SibGAU, 16(3). (2015)
5. N. Ogorodov, A. Kartushinsky, Information technology of satellite data processing for Global Ocean surface gradient characteristics calculation, in CEUR Workshop Proceedings. (2019)
6. A. Kartushinsky, L. Sashko., A. Shevyrnogov, Certificate of state registration of the software. Research of the variability of climatic conditions in the surface layer of the atmosphere. Russia. (2018)
7. J. D. Zika, M. H. England, W. P. Sijp, The ocean circulation in thermohaline coordinates. J. Phys. Oceanogr., 42. (2012)
8. M.G. Lebedeva, A.R. Lupo, Y.G. Chendev, O.V. Krymskaya and A.B. Solovyev, Changes in the Atmospheric Circulation Conditions and Regional Climatic Characteristics in Two Remote Regions Since the Mid-20th Century. Atmosphere, 10, 11. (2019)

Application of remote sensing data to assess environmental situation in Krasnoyarsk

Konstantin Krasnoshchekov[1],* *Oleg* Yakubailik[2]

[1]FRS Ksc SB RAS, Krasnoyarsk, Russia
[2]Institute of Computational Modeling SB RAS, Krasnoyarsk, Russia

Abstract. The work compared the data on ground-based concentrations of suspended particles ($PM_{2.5}$ (particulate matter with diameters less than 2.5 microns) and NO_2 (sulfur dioxide)) with remote sensing data obtained over the territory of Krasnoyarsk in the summer period of 2019 and 2020. We use the air monitoring system of the Krasnoyarsk Scientific Center of the Siberian Branch of the Russian Academy of Sciences (KSC SB RAS) to determine the concentration of PM 2.5. NO_2 concentrations were taken from the data of the State departmental information and analytical system of data on the state of the Ministry of Ecology of the region. It is shown that the data of the MAIAC product, which has a spatial resolution of 1 km, can be used to determine the $PM_{2.5}$ concentration as a supplement to the data obtained by the ground-based air monitoring system of the KSC SB RAS. A high coefficient of determination between satellite and ground monitoring data was obtained. Joint processing of data from ground-based monitoring networks with remote sensing data will contribute to improving the assessment of the ecological situation in Krasnoyarsk.

1 Introduction

The low quality of atmospheric air affects the health of the population and the state of the environment in general. Consequently, the issue of assessing air quality is especially acute in large industrial and developing cities. PM of natural or anthropogenic origin have a significant impact on the climate, the environment and have a negative impact on human health. [1]. Numerous studies have linked an increase in PM concentration with an increase in the number of deaths and an increase in respiratory diseases [2,3].

Toxicology studies [4] have shown that particulate matter 10 μm (PM_{10}) or less has the ability to enter the bloodstream through the lungs, thereby having a negative impact on health. Therefore, air quality assessment, especially in terms of PM_{10} and $PM_{2.5}$, is a pressing issue at the moment.

Nitrogen dioxide (NO_2), like PM, has a negative effect on the body. The main sources of NO_2 are vehicle exhaust gases, emissions from thermal power plants, solid waste incineration, and gas combustion. When a small concentration of nitrogen dioxide enters the respiratory system, a person experiences respiratory failure, coughing, an increase in concentration can lead to oxygen starvation and other negative consequences.

* Corresponding author: krasko@icm.krasn.ru

2 Data and measurements

2.1 Study area

The city of Krasnoyarsk is the regional center of the Krasnoyarsk Territory, with a population of more than 1 million people. It is actively developing and has an area of about 350 km². The coordinates of the city center are 56°00' north latitude and 92°52' east longitude. The city and the surrounding territories have a unique relief. From the south and west of the city, there are forests and hilly terrain. From the north and east, the terrain is mostly flat. The Yenisei River, which does not freeze in winter due to the nearby Krasnoyarsk hydroelectric power station, divides the city approximately in half. Like all major cities, Krasnoyarsk is subject to a negative impact on the environmental situation. Motor transport, the private sector, thermal power plants, large enterprises of the metallurgical, machine-building, and chemical industries contribute many emissions into the atmosphere and negatively affect the quality of the surrounding air. The concentration of $PM_{2.5}$ in Krasnoyarsk is 64% higher than the average in Russia [5].

2.2 Satellite data

To determine the PM concentration using satellite-sensing data, the AOD (aerosol optical depth) parameter is usually used [6]. AOD is the integrated atmospheric dispersion of radiation by aerosols in a vertical column of the atmosphere. This parameter is proportional to the number of particles in the air and depends on their mass concentration.

In our work, we used the data of the MAIAC product. It has a spatial resolution of 1 km. The MCD19A2 (MAIAC) product contains data from the MODIS spectrophotometer installed on the Terra and Aqua satellites. This product was published on May 30, 2018 and contains AOD data from February 1, 2000 [7].

Since data on AOD are available only in the absence of snow cover [7], the period from June to August 2019 and 2020 was taken, a total of 184 days. However, the AOD data is also limited by the presence of cloud cover. As a result, the final number of days used in the calculations was 80.

We used the product OMNO2d as satellite data on nitrogen dioxide. This product provides data for NO_2 in a vertical column of the atmosphere with a spatial resolution of 0.25 by 0.25 degrees. The OMNO2d product contains daily NO_2 values for the entire territory of the Earth from October 1, 2004 to the present [8].

2.3 Ground data

The data on $PM_{2.5}$ of the air monitoring system of the Krasnoyarsk Scientific Center, consisting of 26 monitoring posts, were used in work. A certified CityAir air monitoring station developed by a group of companies from the Novosibirsk Technopark and the Skolkovo Innovation Center is installed at each monitoring post. These stations provide information about the state of the surrounding air, its temperature, pressure, and humidity. Additionally, an optical sensor is installed inside the station to measure the concentrations of $PM_{2.5}$ and PM_{10}. Data on NO_2 were obtained from the State regional environmental departmental information and analytical data system of the Krasnoyarsk Territory. There are 7 automated observation posts (AOP) installed on the territory of Krasnoyarsk, providing information on the concentrations of carbon monoxide, sulfur dioxide, nitrogen dioxide, hydrogen sulfide, and other pollutants. For our study, we used data from two AOPs since data from other AOP had gaps in the study period.

2.4 $PM_{2.5}$ estimation based on satellite data

We have calculated the correlation coefficient between remote sensing data and ground data. Remote sensing data (AOD) has been pre-processed and converted to PM values using formula 1. Based on the value of the calculated correlation coefficient, it is possible to estimate the extent to which satellite monitoring data can be used to assess the environmental situation in the city. For comparison, we took the AOD data located above the ground monitoring station, as shown in figure 1. Then, the obtained values were averaged, obtaining the average value over the city.

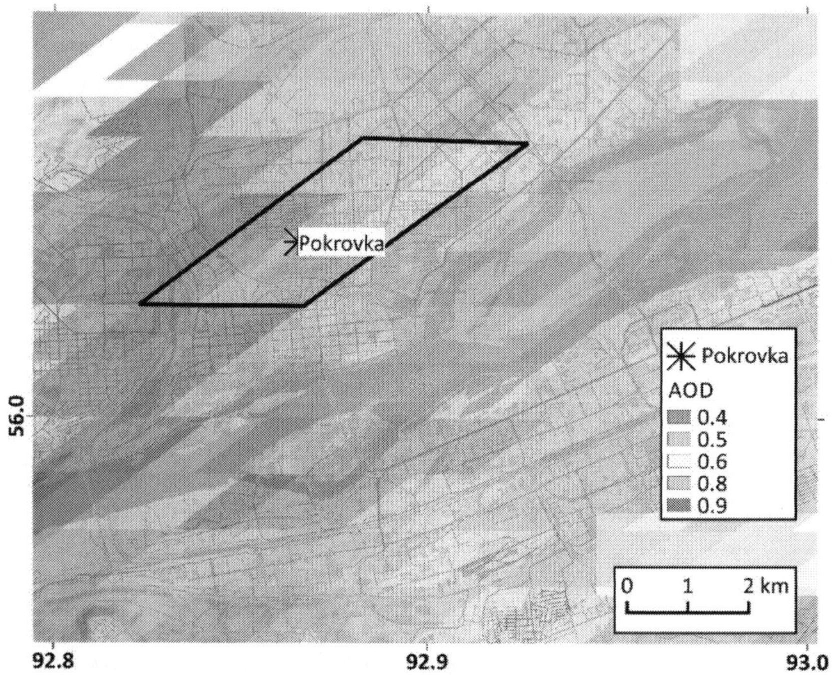

Fig. 1. The area of averaging AOD values above the "Pokrovka" monitoring station.

The data averaging area in figure 1 is shown with a black frame. This area for each station is 3 by 3 pixels (3 by 3 kilometers). This is necessary in order to reduce the amount of missing data. Ground-based monitoring data were taken as the average value of PM2.5 concentration, measured in the period from 12:00 to 13:00 noon, then the data between the posts were averaged to obtain the average $PM_{2.5}$ value in the city.

To compare the AOD and $PM_{2.5}$ data, the AOD data were converted to $PM_{2.5}$ units (μg / m3). We used a model of the kind shown in formula (1). It includes meteorological parameters and takes into account aerosol characteristics. This formula is widely used to compare data on AOD and PM and is described in detail in [5].

$$PM_{calc} = (AOD/PBLH)/\{K \cdot [(1-RH)/(1-RH0)]-\gamma\} \quad (1)$$

To obtain the calculated values according to the formula (1), it is necessary to take into account, in addition to the AOD values, such parameters as: PBLH - the height of the planetaru boundary layer, RH - air humidity, RH0 - the average value of air humidity for the selected area, the variables γ and K are aerosol characteristics and they obtained from [8]. The results obtained by formula (2) are shown in figure 2.

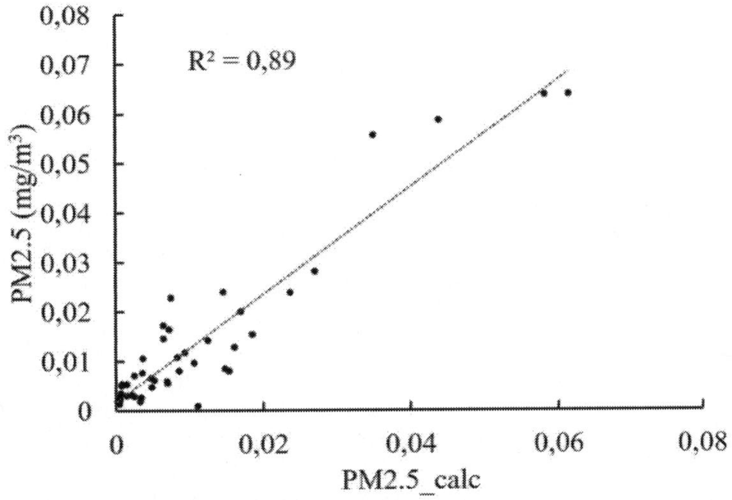

Fig. 2. Comparison of ground data (y-axis) and calculated $PM_{2.5}$ values calculated by formula 1 (x-axis).

By reducing the satellite monitoring data to the units of measurement obtained at ground-based monitoring stations, the obtained two data series were compared and the coefficient of determination between them was calculated ($R^2 = 0.88$). Such a high value of the coefficient of determination indicates that the PM data calculated using satellite monitoring are in high agreement with the data measured at ground-based monitoring stations.

2.5 NO_2 estimation based on satellite data

In order to assess the extent to which satellite monitoring data can be used to assess the environmental situation in the city by the NO_2 parameter, we calculated the correlation coefficient between NO_2 data obtained from ground-based observations and NO_2 data obtained from satellite monitoring data. Since the data has a spatial resolution of 0.25 by 0.25 degrees, the territory of Krasnoyarsk is covered by two pixels, they were averaged among themselves. For comparison of satellite and ground data, the time period was chosen from June 1 to August 31, 2020. NO_2 satellite monitoring data are provided in units of molec/cm³ in the vertical column of the atmosphere, data from ground-based monitoring stations are provided in units of mg/m³. To calculate the correlation coefficient between two data series, we transformed the satellite monitoring data series using formula 2

$$NO_{2\_calc}=a*NO_{2\_sat}+b*RH+c*T+d*P+e*PBLH \qquad (2)$$

where NO_2 sat is NO_2 data obtained using the product OMNO2d, RH is air humidity, T is air temperature, P is air pressure, PBLH is the height of the atmospheric boundary layer. The results obtained using formula 2 are shown in the figure 3.

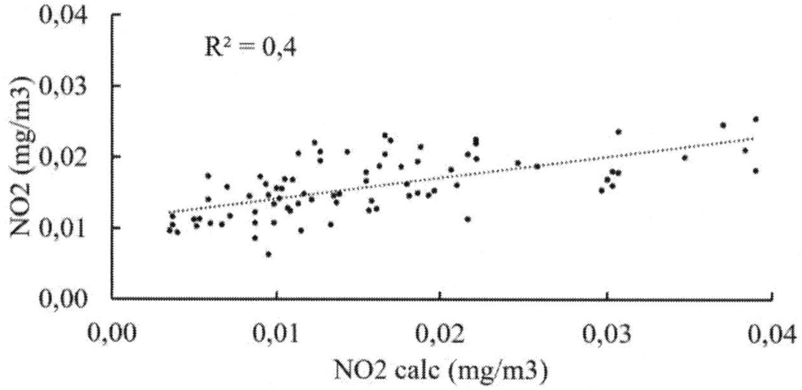

Fig. 3. Comparison of ground data (y-axis) and NO_2 values calculated by formula 2 (x-axis).

The value of the coefficient of determination for NO_2 is worse than the value for $PM_{2.5}$, most likely this is due to the coarser spatial resolution of satellite data. However, for the tasks of assessing the impact of the urban environmental situation on the surrounding suburban area, this accuracy is sufficient.

Using satellite monitoring data, it is possible to significantly supplement the data obtained from ground-based monitoring posts, which will enhance the understanding of atmospheric processes occurring in the city and beyond.

References

1. Y. Kaufman, D. Tanré, O. Boucher, Nature 215–223 (2002)
2. J. Schwartz, F. Laden, A. Zanobetti, Envir. health. perspec. (2002)
3. A. Seaton, D. Godden, W. MacNee, K. Donaldson, The Lancet **345** 176–178 (1995)
4. B. Brunekreef, S. Holgate, The Lancet **360** 1233–1242 (2002)
5 C. Lin, L. Labzovskii, H. Mak, J. Fung, A. Lau, S. Kenea, M. Bilal, J. Vande Hey, X. Lu, J. Ma, Atmos. Environ. **227** (2020)
6. K. Donaldson, X. Li, W. MacNee, J. Aerosol Sci. **29** 553–560 (1998)
7. A. Lyapustin, Y. Wang, S. Korkin, D. Huang, Atmos. Measur. Techniq. **11** 5741–5765 (2018)
8. Team O.M.I.N.O.A http://disc.sci.gsfc.nasa.gov/Aura/data-holdings/OMI/documents/v003/OMNO2d_FileSpec **3** (2013)

Analysis and assessment of agroclimatic conditions of cultivation and yield of technical crops on the territory of Armenia

Varduhi Margaryan [1,*], *Gennady* Tsibulskii[2] and *Ksenia* Raevich[2]
[1]Yerevan State University, Department of Physical Geography and Hydrometeorology, Faculty of Geography and Geology, Yerevan, Armenia
[2] Siberian Federal University, Institute of Space and Information Technologies, Chair of Systems of Artificial Intelligence, 660074 Kirenskogo St. 26, Krasnoyarsk, Russian Federation

Abstract. The necessary agroclimatic conditions for the growth and development of technical crops on the territory of the republic, the terms of the phenological phases, the prerequisites for the development of their cultivation space, the level of yield and crop yields have beenanalyzed and evaluated. As a source of information, the actual data of agrometeorological observations of the GEO «Center for Hydrometeorology and Monitoring» of the Ministry of Environment of the Republic of Armenia and the National Statistical Service are used. For the period 2012-2020, there is a tendency to reduce the acreage of technical crops. So that the Republic of Armenia is not only a country importing technical crops, but also an exporting country, apply modern high technologies of cultivation, expand irrigated areas, introduce new high-yielding varieties, pay special attention to more efficient use of microclimatic resources.

1 Introduction

The role and significance of the analysis and assessment of the agroclimatic conditions of agricultural crops from the point of view of the prospective development of agriculture, food and light industry is very important. They determine the role of heat supply and moisture supply of a given territory, growth and development of crops, conditions for their wintering, timing of field work, yield.

Agriculture is the most dependent on climate among the industrial sectors of the economy: it almost always suffers from weather changes, dangerous agrometeorological events and natural disasters. Food security risks are increasing as a result of expected climate change. Therefore, the strategic objective of the agriculture of the Republic of Armenia is to increase the food security of the republic, ensuring 75-80 % of its own production of basic food.

One of the issues with respect to climate change involves its influence on the distribution of future crop yields [6]. Projections of future climate change suggest that average crop yield will increase with warmer temperatures and a longer growing season which is only partially offset by forecast increases in the variability of temperature and rainfall. The projections would also depend on future technological developments, which have generated significant increases in yield over time despite changing annual weather conditions [5]. The estimation results indicate that changes in climate modify crop yield levels and variances in a cropspecific fashion. For sorghum, rainfall and temperature increases are found to increase yield level and variability. On the other hand, precipitation and temperature are individually found to have opposite effects on corn yield levels and variability [6].

Technical are considered those crops that are raw materials for production or are used after technical processing. The following groups of industrial crops are distinguished: sugar (sugar beet, sugar cane), oil (sunflower, soy, hemp, peanuts, rapeseed, sesame, flax, olive, oil and coconut palm, etc.), fibrous (cotton, fiber, etc.), invigorating (tea, coffee, cocoa), essential oil (geranium, rose, anisson, mint, cumin, lavender, etc.), rubbery (hevyea), tanning (valerian, ginseng, hawthorn, etc.), core plants (Amur velvet wood, etc.). Some types are used for different purposes.

Until 1988 sugar beet (Shirak and Lori regions), geranium (Armavir region) were cultivated in the republic, until 1965 – cotton (Ararat valley) [7].

The purpose of the work is to evaluate the necessary agroclimatic conditions for the growth of industrial crops on the territory of the republic, the timing of the phenological phases, the areas of their cultivation, the tendency of the development of the level of crop yield and yield.

* Corresponding author: vmargaryan@ysu.am

2 Study areas, Data and Methods

To solve these problems in the work as a theoretical and informational basis were the relevant research studies [1, 3, 8-9]. As a source material, the work used actual agrometeorological observations of the «Hydrometeorology and Monitoring Center» SNCO of the Ministry of Environmental Protection of the Republic of Armenia, as well as data from the NSS of RA, agroclimatic reference books and yearbooks [2-4].

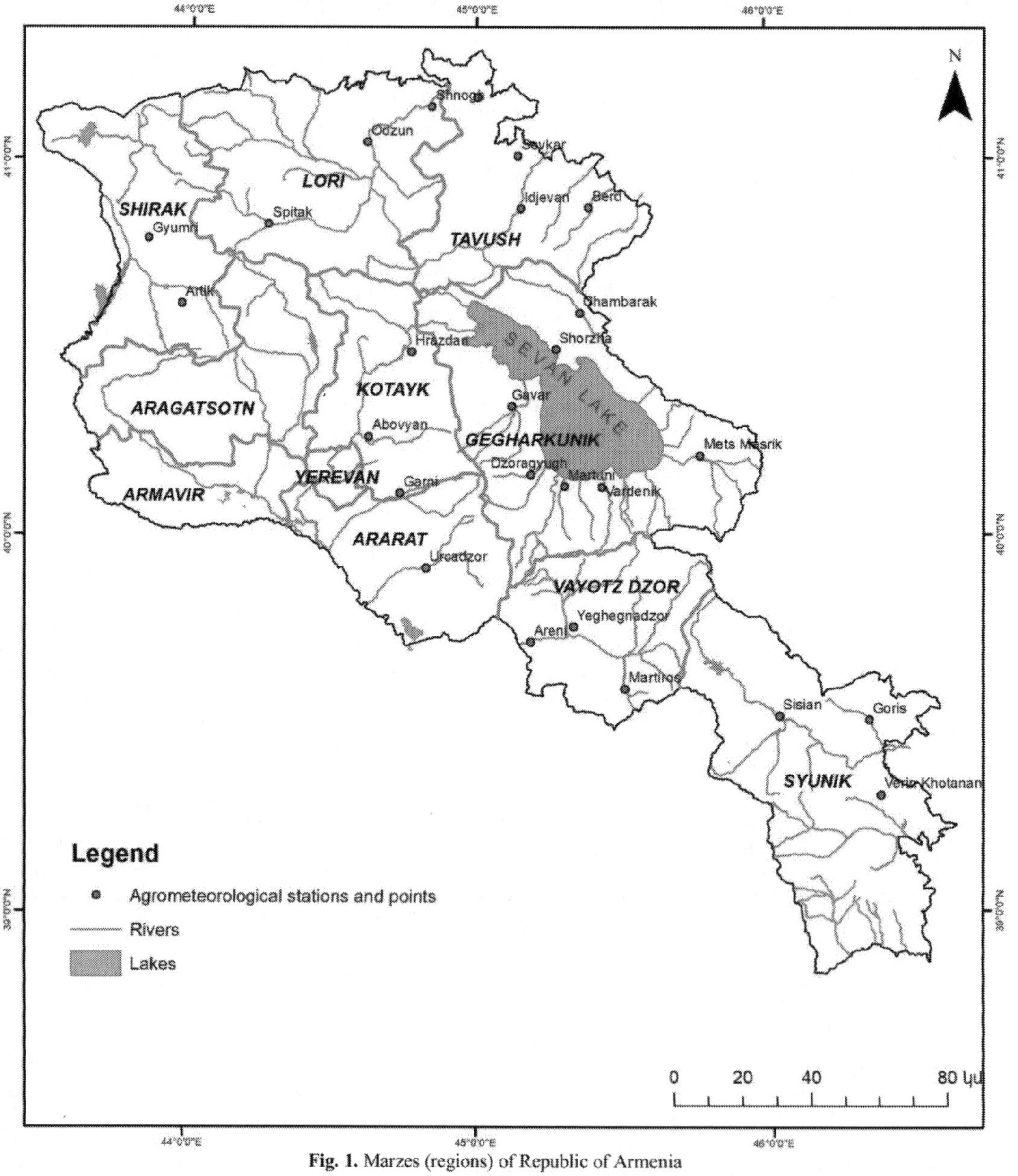

Fig. 1. Marzes (regions) of Republic of Armenia

At the meteorological stations and posts of Armenia, observations were carried out for a long time on the development periods of the phases of crops cultivated in a given area (that is, phenological observations were organized) [7]. In the republic from industrial crops phenological observations are carried out tobacco. In tobacco: 3rd leaf, 5th leaf, 7th leaf, steam growth, the appearance of the first stepsons, the appearance of inflorescences, flowering, technical ripeness of the leaves of the lower tier, the timing of leaf collection, as well as damage to plants due to adverse and dangerous phenomena (frost, hail, diseases, pests, etc.) during the growing season, the likelihood of frost damage.

It should be noted that the number of those stations that conduct phenological observations of industrial crops has sharply decreased (fig. 1). So, since 2000 phenological observations only of tobacco have been carried out in Urtsadzor.

Only industrial crops grown in the republic are discussed in the work.

As a methodological basis used are: statistical analysis, correlation methods.

3 Results and discussion

According to the results of actual observations of the «Hydrometeorology and monitoring center» SNCO, table 1 shows the average sowing time and development phases of sugar beet, and table 2 – agroclimatic conditions of the sowing-sprouting period at different sowing dates.

Table 1. Average sowing time and development phases of sugar beet [4]

Meteorological Stations	Sowing	Seedlings	5th true leaf	Thickening of the hypocotal knee
		Lori		
Spitak	02/05	16/05	11/06	25/06
		Shirak		
Gyumri	28/04	15/05	05/06	18/06
Artik	04/05	24/05	15/06	30/06

According to table 1, phenological observations of sugar beet were carried out on the territory of the republic at the meteorological stations Spitak, Gyumri and Artik. Sowing is mainly carried out from the third decade of April to the first decade of May.

Tobacco is an annual plant. It is grown mainly for the purpose of obtaining raw materials for smoking. The growing season of a tobacco plant is divided into two periods: seedling – from sowing seeds to plants ready for seedling, and field – from seedlings to the final collection of leaves or seed ripening. The duration of the planting period is 35-50 days. The best temperature for germination of tobacco seeds is 25,0-28,0 °C, and in the future – 18,0-22,0 °C. Tobacco is sensitive to late spring and early autumn frosts, it is damaged of temperatures from -1,0 to -3,0 °C. The best period for transferring tobacco seedlings to the field is when the temperature in the soil reaches 10,0-12,0 °C. Tobacco must be watered with grooves, 3-10 times, according to the needs of the culture [7].

Sugar beet (Bets Vulgaris L. v. saccharifera) – biennial plant. It has a wide range of consumption. When processed, sugar is obtained, which can be used as a final or as an average product. Currently about 40% of the sugar used in the world is obtained from sugar. In the process of sugar production, sugar beet kernels are obtained from sugar beet, which are a complete feed for animals. Sugar beet is also processed as a valuable juicy food. Sugar beet leaves are used as animal feed or mixed with soil to increase fertility [10]. In Armenia, the only large enterprise for the production of sugar was «Lus Astkh sugar», the Akhuryan sugar factory, after the operation of which (2010) in the republic there was a need to resume the production of sugar beet.

Sugar-beet – heat-loving, light-loving, moisture-loving, salt-tolerant and drought-tolerant technical crop. Sugar beet seedlings may be damaged due to frost in Gyumri, the probability of frost is 4-8% (table 2). In Artik and Spitak the probability of freezing of sugar beet seedlings is 0 %.

Table 2. Agroclimatic conditions of sugar beet during the sowing-germination period at different sowing dates [4]

Meteorological Stations	Sowing	The probability of the condition of the seed in the soil, %			Germination date	The probability of freezing of seedlings, %
		the temperature at a depth of 5 sm will be below 7°	the soil will be dry	the soil will be waterlogged		
			Shirak			
Gyumri	11/04	61	20	16	02/05	8
	21/04	45	25	20	06/05	8
	01/05	7	33	20	13/05	4
	11/05	0	25	25	21/05	4
	21/05	0	29	20	28/05	0
Artik	11/04	58	8	0	04/05	0
	21/04	58	8	17	08/05	0
	01/05	17	17	8	14/05	0
	11/05	0	17	17	24/05	0
	21/05	0	8	25	31/05	0
			Lori			
Spitak	11/04	36	36	4	03/05	0
	21/04	12	34	13	08/05	0
	01/05	0	33	18	14/05	0
	11/05	0	17	4	21/05	0
	21/05	0	31	5	31/05	0

Unlike sugar beet, phenological observations of tobacco on the territory of the republic were organized at many meteorological stations. Planting of seedlings is mainly carried out from the first decade of May to the first decade of June, and the collection of leaves – from July-August (first collection of leaves) to September-October (last collection of leaves). With height, there is a delay in the average timing of development of tobacco phases (table 3).

Table 3. Average periods of tobacco development phases [4]

Meteorological Stations	Variety	Transplanting	The appearance of inflorescences	Flowering	Leaf picking time	
					first	last
Ararat						
Urtsadzor	Ostrolist 2747	08/05	06/08	14/08	17/07	27/09
Gegharkunik						
Chambarak	Samsun 935	03/06	18/08	31/08	31/07	23/09
Shorzha	Ostrolist 2747	26/05	30/08	19/09		
Masrik	Samsun 36	08/06	13/09	23/09	02/08	23/09
Martuni	Samsun 935	12/06	26/08	05/09	02/08	07/10
Gavar	Samsun 935	08/06	27/08	07/09	28/07	25/09
Vardenik	Samsun 935	25/05	25/08	31/08	30/07	02/10
Dzoragyugh	Samsun 935	24/05	22/08	30/08	28/07	28/09
Lori						
Shnogh	Ostrolist 2747	09/05	19/07	27/07	05/07	17/10
Odzun	Trapezond	15/05	06/08	14/08	03/08	09/10
Kotayk						
Hrazdan	Samsun 935	05/06	05/09	14/09	06/08	27/09
Abovyan	Samsun 36	20/05	15/08	24/08	23/07	30/09
Garni	Samsun 935	21/05	15/08	31/08	25/07	01/10
Syunik						
Goris	Samsun 935	15.05	26/07	02/08	20/08	24/09
Khotanan	Samsun 935	13.05	16/08	29/08	16/07	08/09
Sisian	Samsun 935	27.05	25/08	04/09	02/08	23/09
Vayots Dzor						
Areni	Samsun 935	12/05	08/08	26/08	12/07	06/10
Eghegnadzor	Samsun 935	08/05	30/07	08/08	14/07	02/10
Martiros	Samsun 935	31/05	02/08	06/09	30/07	30/09
Tavush						
Ijevan	Ostrolist 2747	13/05	02/08	10/08	23/07	30/09
Noyemberyan	Ostrolist 2747	14/05	12/08	23/08	20/07	25/10
Berd	Peremozhets 83	13/05	16/08	25/08	13/07	12/10
Sevkar	Ostrolist 2747	19/05	04/08	15/08	15/07	07/10

According to the data of the National Statistical Service, table 4 shows the area of cultivation, gross yield and trends in the development of growth in the yield of the currently cultivated main technical crops (tobacco, flax, sunflower) in the territory of Armenia for 2010-2020. It turns out that among technical crops, the largest sown area on the territory of the republic is occupied by sunflower. Information on the sown area, gross yield and yield of sunflower is contained for 2015-2020.

Table 4. Area of cultivation of tobacco, flax and sunflower, gross yield and growth trends in yield in Armenia for 2010-2020

Characteristics	Years										
	2010	2011	2012	2013	2014	2015	2016	2017	2018	2019	2020
Tobacco											
Sown area, ha	418	614	667	548	510	582	464	460	393	381	283
Harvested crop, centner	13758	23862	18170	16784	12528	15157	11321	10867	9140	9466	7158
Average crop capacity of hectare, center	33,0	38,9	27,2	30,6	24,6	26,0	24,4	23,6	23,3	24,8	25,3
Flax											
Sown area, ha	67	52	38	62	90	154	202	267	325	174	192
Harvested crop, centner	823	550	555	824	886	1785	2549	4718	5456	1396	2292

Average crop capacity of hectare, center	12,3	10,6	15,0	13,5	10,0	12,9	12,7	17,7	16.8	8,0	11,9
Sunflower (oil and confectionery)											
Sown area, ha	-	-	-	-	-	1976	1819	1353	863	797	601
Harvested crop, centner	-	-	-	-	-	31799	29262	22053	15365	21258	18240
Average crop capacity of hectare, center	-	-	-	-	-	17,4	16,1	16,7	17,8	26,7	30,3

The territories sown with industrial crops, including tobacco and flax, increased over the period 2005-2020 (Fig. 2a, 2b). Moreover, the sowing of tobacco for the period 2012-2020 decreased. In general, in recent years, there has been a tendency for a decrease in the area sown with industrial crops (tobacco, green, sunflower). Therefore, it is necessary to take appropriate measures and increase the sown area of industrial crops in the republic.

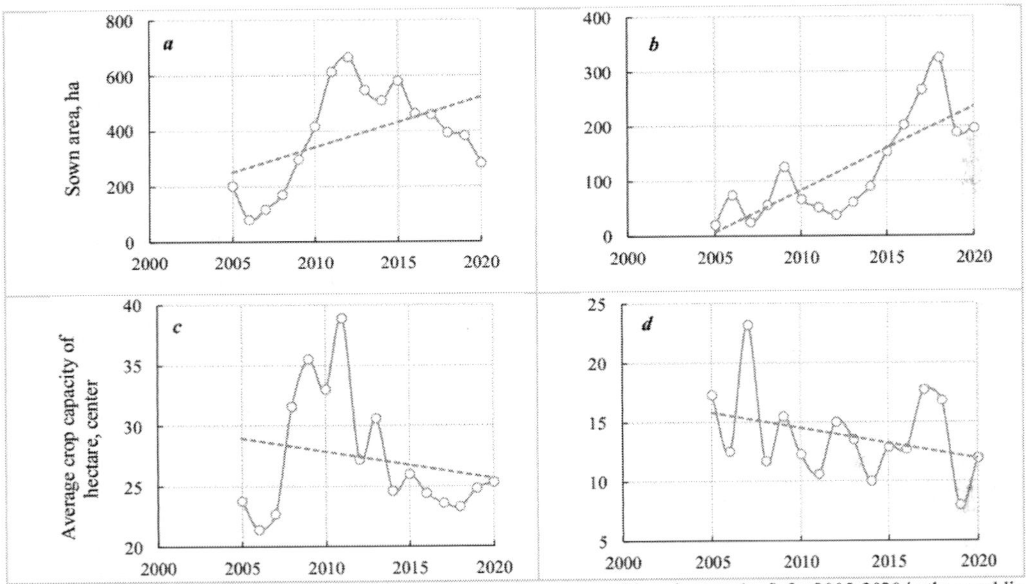

Fig. 2. Changes in cultivated areas of industrial crops (a, b) and average yield per hectare (c, d) for 2005-2020 in the republic. Tobacco - a, c, flax - b, d.

On the other hand, for the period 2005-2020, there is a tendency to decrease the yield of tobacco and flax (Fig. 2c, 2d).

Average crop yields increase at a decreasing rate with the quantity of inputs used, and decrease with the area planted to the crop. Climate variables have a major impact on mean yield with the length of the growing season being the primary determinant. Increases in the variability of temperature and precipitation decrease mean yield and increase its variance. Yield variance is poorly explained by both seasonal and monthly climate variable models [5].

4. Conclusions

So,

✓ Agroclimatic conditions of the territory of the republic are favorable for the growth and development of technical crops.

✓ On the territory of the republic there is tendency to crease the acreage of tobacco and flax, decrease their yield for the period 2005-2020.

✓ On the territory of the republic, in general, for the period 2012-2020, there is a tendency to reduce the acreage of technical crops.

It is necessary:

✓ So that the Republic of Armenia is not only a country importing technical crops, but also an exporting country.

✓ Apply modern high technologies of cultivation, expand irrigated areas, introduce new high-yielding varieties.

✓ Pay special attention to more efficient use of microclimatic resources.

References

1. Agriculture sustainability and risk assessment under climate change. Global climate change and risk assessment in agriculture in Russia. (Saint-Petersburg. 2011)
2. Agroclimatic resources of Armenia // R.S. Mkrtchyan. Leningrad, Gidrometeoizdat, 1961. 266 p. (In Russian)
3. Agroclimatic resources of Armenia // R.S. Mkrtchyan. Leningrad, Gidrometeoizdat, 1976. 388 p. (In Russian)
4. Agroclimatic resources of Armenia // Mkrtchyan R.S., Melkonian D.H., Badalyan V.H. MES of Armenia, Hydrometeorological and Monitoring Service of Armenia. Yerenan, 2011. 155 p. (In Armenian).
5. Cabas J., Weersink A., Olale E.Crop yield response to economic, site and climatic variables // Climatic Change (2010) 101:599–616 DOI 10.1007/s10584-009-9754-4
6. Chen C.C., McCarl, B.A., Schimmelpfennig, D.E. (2004). Yield variability as influenced by climate: A statistical investigation. Climatic Change, 66(1–2), 239–261.
7. Margaryan V.G., Aleksanyan K.B., Muradyan Z.Z. Assessment of the agroclimatic prerequisites for the cultivation of industrial crops (on the example of RA) // Proceedings of the YSU C: Geol. Geogr. Sci. 2021, 55, 120-128. DOI: https://doi.org/10.46991/PYSU:C/2021.55.2.120. (In Armenian).
8. Margaryan V., Tsibulskii G., Fedotova E.Impact of climate change on agricultural crops yield on the territory of the Republic of Armenia // E3S Web of Conferences, Volume 223 (2020), id. 03019. Regional Problems of Earth Remote Sensing (RPERS 2020), 2020. https://doi.org/10.1051/e3sconf/202022303019
9. Martsinevskaya LV, Sazonova NV, Soloviev A.B. Agroclimatic conditions and yield of industrial crops in the Belgorod region // Scientific bulletin of BelSU. Belgorod, 2015. Pp. 260-264.
10. Rafati Mohsen The estimation of return of research works investment in sugar beet breeding (by the example of Islamic Republic of Iran). The thesis abstract for the degree of candidate of economic sciences. Yerevan. 2017. 32 p.

FEATURES OF THE MAXIMUM WATER FLOWS OF THE ARPA RIVER UNDER MODERN CONDITIONS

Varduhi Margaryan[1,*], *Levon* Azizyan[2], *Amalya* Misakyan[2], *Ekaterina* Gaidukova[3], *Gennady* Tsibul'skii[4], *Ksenia* Raevich[4]

[1] Yerevan State University, Department of Physical Geography and Hydrometeorology, Faculty of Geography and Geology, 0025, Alek Manoukian 1 st., Yerevan, Armenia

[2] Hydrometeorology and Monitoring Center SNCO with Ministry of Environment of the Republic of Armenia, Yerevan, Armenia

[3] Russian State Hydrometeorological University, Saint-Petersburg, Russia

[4] Siberian Federal University, Institute of Space and Information Technologies, Chair of Systems of Artificial Intelligence, 660074 Kirenskogo St. 26, Krasnoyarsk, Russia

Abstract. The paper discusses the main regularities of the peak flood discharge distribution in modern conditions, using actual data of Hydrometeorology and Monitoring Center SNCO with Ministry of Environment of the Republic of Armenia on the peak flood discharge of the river Arpa.

1 Introduction

Water management planning, assessment of water resources, and construction design of hydraulic structures should take account of the peculiarities of the water regime formation in rivers, water content, maximum and minimum runoff, which contributes to the relevance of alternative methodological approaches to assessing the extreme characteristics of runoff, in particular, the peak flood discharge of unexplored rivers [9]. The maximum water discharge during the flood period is one of the most important characteristics of this phase of the water regime. Meaning maximum flow rate is very important for economic activities, as it determines the possibility of flooding adjacent to the watercourse territories [4]. As one of the most common natural phenomena, floods can bring both risks and benefits for human beings. They can pose a risk of inundation to a human habitat but can also be utilized as a resource with hydraulic engineering. Improving the knowledge of flood characteristics is the basis and premise of improving water resources management and ecological environmental protection [17]. Determination of the peak river water discharge is a mandatory stage of engineering surveys for the construction of all water management systems, bridges, pipeline crossings and a number of other structures [13]. Significant influence of anthropogenic factors (reservoirs) on the maximum flow of the spring flood allow us to estimate the changes of values of the maximum flow in number [1].

That is, the role of knowledge about peak discharge is great, especially from the point of view of the hydraulic structure design and operation, and implementation of measures to prevent the damage caused. Peak discharge rates are one of the most important characteristics of the flood phase of the water regime.

The purpose of this article is to analyze and assessing the features of peak discharge distribution in the current conditions of the Arpa river.

Studying the issues of space-time distribution, formation and forecasting of maximum water discharge in rivers At one time, many Armenian scientists-hydrologists were engaged in Armenia, in in particular, M.V. Shahinyan [14], V.G. Margaryan and others [7–8], A.E. Misakyan and others [10], based on the results of many years of research, monographs [3, 16] etc.

2 Study areas, Data and Methods

The Arpa River is a large left tributary of the river. Araksa: along its length it the fourth in the Republic of Armenia (RA) [8]. The length of the Arpa River is 128 km, total the catchment area is 2630 km^2, but on the territory of the Republic of Armenia there are 92 km and 2080 km^2, respectively (of which 1880 km^2 is the catchment area post of Areni). The Arpa River originates from the northwest slope Syunik highlands from a height of 3260 m (fig. 1). The estuary is at a height 960 m. The river in the upper and middle reaches, as well as its tributaries, have features

* Corresponding author: vmargaryan@ysu.am

inherent in mountain rivers: pass through large gorges and valleys cut by tributaries and ravines. Average slope of the catchment the basin of the Jermuk post is 188 ‰ (table 1). The studied territory, the average height of the catchment area at water measuring posts is in range from 2110 to 2790 m, and the catchment area - from 199 to 1880 km^2.

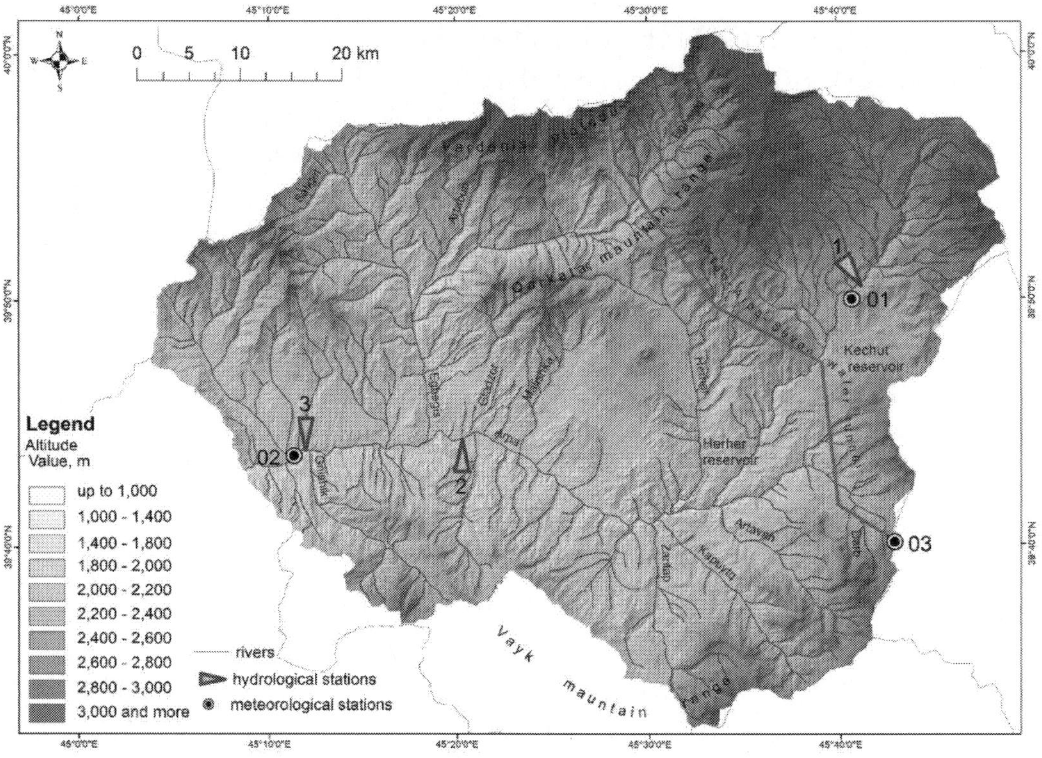

Fig. 1. Location of hydrological posts in the river Arpa and meteorological stations in the there basin
hydrological posts (p.): 1 – r. Arpa – p. Jermuk, 2 – r. Arpa – p. Ekhegnadzor, 3 – r. Arpa – p. Areni;
meteorological stations: 01 – Jermuk, 2 – Areni, 03 – Vorotan Pass.

The basin is composed mainly of volcanic rocks. The highest parts of the mountains are composed of andesite-basalts, below are widespread received andesites, tuffs, tuff breccias and other rocks. Below p. Areni Arpa goes into the area of foothills, where the volcanic stratum is replaced dislocated sedimentary rocks - limestones, sandstones, schists and quartzites of the Paleozoic. Gradually lowering towards the south and southeast, the foothills pass into the Priaraksa lowland (800-900 m) [11].

Table 1. Main hydrometric and hydrological characteristics of the Arpa River basin rivers and their catchments [8]

№	River – post	Distance from the mouth, km	Slope of river, ‰		Main characteristics of the catchment			Average annual water flow, m^3/s
			average from the most distant point	weighted mean from the most distant poin	area, km^2	average height, m	average slope, ‰	
1	Arpa – Jermuk	105	52	48	199	2790	188	5.15
2	Arpa – Ekhegnadzor	56	30	22	1220	2140	–	7.75
3	Arpa – Areni	40	26	20	1880	2110	–	13.6

The waters of the river are used for irrigation and hydropower purposes. A significant part of its waters through the Vorotan-Arpa-Sevan tunnel flows into the lake. Sevan. The long-term average water consumption at the closing post of Areni is 13.6 m^3 / s for the period 1981–2019 (table 1).

The main role in the formation of the maximum flow rates of the river. Arpa plays the amount of snow storage and heat inflow. Amount and date of occurrence maximum water consumption is associated with the amount of snow reserves, the intensity of the inflow of warm air and precipitation. Average annual air temperature according to the meteorological station is in range from 3.1 to 12.9 °C, annual precipitation - in the range from 374 up to 764 mm (table 2). The maximum air temperature can be up to +42.6 °C, and the minimum - up to –30.0 °C. Relative humidity ranges from 60 to 80%, and the lack of air humidity - from 2.2 to 9.3 hPa.

Table 2. Average long-term actual values of meteorological characteristics for the period 1981–2019 at meteorological stations of the Arpa River basin

Meteorological stations		Jermuk	Areni	Vorotan Pass
Air temperature, °C	average	5.1	12.9	3.1
	absolute maximum	33.8	42.6	30.2
	absolute minimum	-30.0	-22.4	-27.5
Atmospheric precipitation amount, mm		764	374	619
Humidity	relative, %	71	60	80
	deficit, hPa	3.81	9.32	2.24

The theoretical and informational background were the results presented in the work [2, 5–6, 12, 15]. Actual observations of Hydrometeorology and Monitoring Center SNCO with Ministry of Environment of the Republic of Armenia for the years 1957-2020 were used as the initial material. The peak discharge was studied in those section lines of the rivers that are located directly on the Arpa river (fig. 1). Actual observations of the peak discharge were carried out since the early 30s of the 20th century in Yeghegnadzor (about 1075 m) and Areni (about 999 m) section lines located in the lower reaches of the Arpa river, and since 1957 in Jermuk section line, which is located upstream (about 2034 m). Therefore, for the purpose of comparison and referencing, features of peak discharge of the Arpa river in modern conditions were studied in the work for the period of 1957-2020. However, in 1981, operation of the Kechut reservoir started, which was built on the Arpa river, downstream from Jermuk section. Accordingly, to assess the impact of the reservoir on the peak discharge in Yeghegnadzor and Areni section lines of the Arpa river, two periods were distinguished: before the reservoir operation (1957-1980) and after the operation (1981-2020).

This study applies the method of mathematical and statistical analysis, the method of comparison and referencing, extrapolation and correlation, and complex geographic-graphical-hydrometeorological analysis.

3 Results and discussion

Due to the variety of natural conditions of the studied area the absolute maximum flow rates of the spring flood are distributed very unevenly (table 3), but the leading role belongs to climate in combination with the features of the relief, soil and vegetation cover and geological and hydrogeological structure of river catchments. At volume, conditions for the formation of the maximum water consumption from year to year vary depending on meteorological factors. Absolute maximum flow rates of the river. Arpa range from 91 m^3 / s to 199 m^3/s for the period 1980-2020.

In the study area, the absolute peak water discharge occurs mainly during the spring flood. In the period from 1957 to 2020, the absolute peak water discharge in Jermuk section line of the Arpa river was observed during the spring flood in all years, while in Yeghegnadzor and Areni section lines of the Arpa river is was observed in 1.6% of cases. During the flood period along the rivers of the river basin. Arpa on average passes about 55–62% of the total runoff per year. Average volume of spring flood r. Arpa varies widely: from 95.5 million m^3 (r. Arpa – p. Jermuk) up to 277 million m^3 (r. Arpa – p. Areni). Spring flood looks like a well-defined wave formed by melted snow, rain and groundwater.

Table 3. The main long-term characteristics of the peak flood discharge of the Arpa river for the periods before and after operation of the reservoir

River – post	Observation period		Long-term run-off characteristics						
			Maximum for the observation period and the year			Over a long period			
	Years	Number of years	Discharge, m^3/s	Runoff module, $l/s \cdot km^2$	Day, month, year	Average discharge, m3/s	Variation coefficient, Cv	Asymmetry coefficient, Cs	Cs/Cv
Arpa – Jermuk	1957-2020	64	91,0	457	17/05/1983	46.0	0.40	- 0.30	-0.75
Arpa – Yeghegnadzor	1957-1980	24	244	200	29/04/1969	106	0.38	1.72	4.53
	1981-2020	40	131	107	29/04/1997	64.6	0.40	0.56	1.40
	difference		113	93		41.4			
Arpa – Areni	1957-1980	24	340	181	01/04/1969	168	0.43	1.05	2.44
	1981-2020	40	199	106	18/04/1988	115	0.40	0.04	0.10
	difference		141	75		53			

Downstream of the Arpa river, the peak discharge values increase and reach large values in the outlet (таble 3). The highest peak discharge, observed on 18 April 1988 in Areni settlement section line, was 199 m^3/s, and before the construction of Kechut reservoir, on 1 April 1969, the river passed 340 m^3/s of water. After construction of the reservoir, the flow of the river was regulated. As a result, for the period of 1981-2020 there was a decrease in the main characteristics of the peak spring flow discharge as compared with the period of 1957-1980 (fig. 2). So, the Danjiangkou Reservoir has changed the flood characteristics to a great extent both of seasonal or annual floods, and the mean degree of the alteration is about 19 % [17].

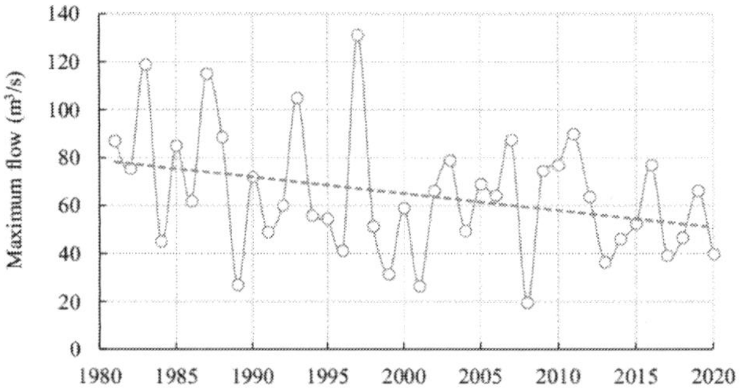

Fig. 2. Dynamics of changes in maximum flow on Arpa river – Yeghegnadzor post

Table 3 shows that the variation coefficient (Cv) of the absolute peak discharge of the Arpa rivers does not exceed 0.50 for the entire period under discussion and changes within a small range from 0.46 to 0.40. Such variation values are rather small for absolute peak flow discharge rates. That is, it can be assumed that the absolute peak discharges of the Arpa rivers are quite stable and have relatively low risk values. Unlike the variation coefficient, the series asymmetry coefficient (Cs) has large values varying from 1.05 to 1.72 for the period 1957-1980 and from 0.04 to 0.56. The asymmetry coefficient shows distribution of the series members in relation to the central member of the series, if the series is presented in decreasing or increasing order.

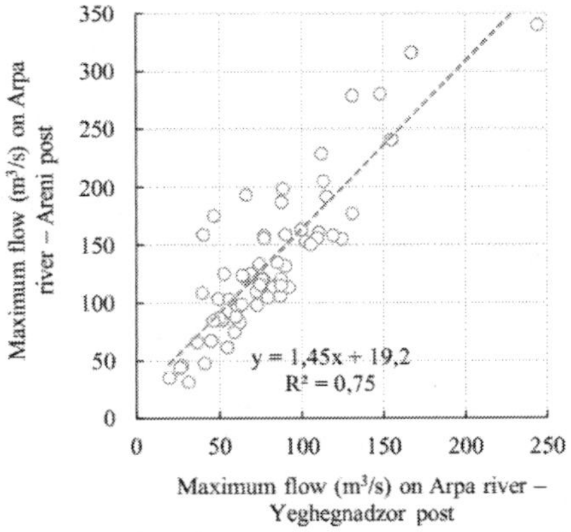

Fig. 3. Correlation between the absolute maximum flow (m^3/s) values on Arpa river – Yeghegnadzor post and Arpa river – Areni post

A correlation is noted between the values of the absolute peak discharge values for the flood period (IV-VI) on the Arpa rivers in Areni and Yeghegnadzor section lines. A dependence of this kind can be used to calculate and forecast the maximum runoff of the spring flood (Fig. 3).

4. Conclusions

Thus, the absolute maximum flow occurs during the spring flood, in Jermuk section line of the Arpa river. The absolute maximum flow values were observed during the spring flood in all years in the period from 1957 to 2020. In Yeghegnadzor and Areni section lines they were observed in 1.6% of cases.

Downstream of the Arpa river, the maximum flow values increase to reach high values in Areni outlet.

In the period of 1981-2020, there is a decrease in the main characteristics of the peak flood discharge as compared with the period of 1957-1980.

The series variation coefficient of the absolute peak discharge values varies within a small range, from 0.46 to 0.40; the asymmetry coefficient varies in a relatively large range of 1.05 to 1.72 for the period of 1957-1980 for Yeghegnadzor and Areni section lines of the Arpa river.

A correlation has been revealed between the absolute peak discharge values for the flood period on the Arpa river in Yeghegnadzor and Areni section lines, which can be used for calculating and forecasting the peak flood run-off.

Financing. This work was supported by the RA Science Committee and Russian Foundation for Basic Research (RF) in the frames of the joint research project SCS 20RF-039 Short-term probabilistic forecast of river flow during the spring flood and RFBR № 20-55-05006\20 accordingly.

References

1. Arystambekova D., Jussupbekov D., Skakova A., Kujibayeva G., Mamirova K. Assessment of Changes in the Maximum Flow of Kazakhstan Plain Rivers // Journal of Ecological Engineering Vol. 20(5), 2019
2. Gaidukova E.V., Margaryan V.G., Myakisheva N.V., Pavlov M.R. and Khaustov V.A. Review of methods for effective forecasting of river runoff characteristics in mountain and semi-mountain areas // IOP Conference Series: Earth and Environmental Science, Volume 867, 012006. International Symposium «Earth sciences: history, contemporary issues and prospects» 10 March 2021, Moscow, Russian Federation. doi:10.1088/1755-1315/867/1/012006
3. Hydrography of ASSR. Yerevan: Academy of Sciences of ASSR, 1981: 177 p. (In Armenian).
4. Kireyeva M.B., Frolova N.L. Present-day special features of the Don river basin rivers spring tide // Water Sector of Russia: Problems, Technologies, Management. 2013. No. 1. Pp. 60–76. (In Russian).
5. Komlev A.M., Melnikova T.N. Water regime of the rivers of the North-West Caucasus. Formation patterns and calculation methods. Perm: PSU, 2008.112 p. (In Russian)
6. Kuzmin V.A., Gavrilov I.S., Sokolova T.A., Timofeev A.Yu. Calculation of the peak flood discharge by the method of integral correction coefficient optimization // Scientific notes of the Russian State Hydrometeorological University, 2010. No. 14, pp. 5–13. (In Russian)
7. Margaryan V.G. The regularities of temporary changes in runoff in the high water period of the Arpa river basin in the context of stable development // Geosphere Research. 2019. № 1. Pp. 44–53. DOI: 10.17223/25421379/9/3. (In Russian).

8. Margaryan V.G., Gaidukova E.V., Azizyan L.V., Misakyan A.E. Specific Features of the Spring Flood Formation in the Arpa River Basin // Water Sector of Russia: Problems, Technologies, Management. 2021. No. 3. P. 126-152. DOI: 10.35567/1999-4508-2021-3-7. (In Russian)
9. Melnikova T.N. Maximum runoff of the spring flood of the rivers of the North-West Caucasus // Vestnik ASU, 2014. Vol. 3 (142). pp. 158-164. (In Russian)
10. Misakyan A.E., Azizyan L.V., Azizyan H.H. Forecast of maximum discharge for the Jermuk observation post on the Arpa river // Modern problems of hydrology, meteorology and climatology in Armenia. Materials of the scientific seminar detieated to World Water and Meteorology Day, March 21-22, 2014. Yerevan: Pub. house «Lusabats», 2014. Pp. 70-73. (In Armenian).
11. Physical Geography of the Armenian SSR. Yerevan. Publishing House of the Academy of Sciences of Armenia. USSR, 470 p. (Armenian).
12. Razmi Ali G.S., Zahmatkesh Z. Non-Stationary Frequency Analysis of Extreme Water Level: Ap-plication of Annual Maximum Series and Peak-over Threshold Approaches. Water Resour. Manag. 2017, 31, 2065–2083. DOI: 10.1007/s11269-017-1619-4
13. Savichev O.G. Methodology for calculating the maximum discharge of river waters in the taiga zone of Western Siberia // Izvestia TPU, 2011. V. 318, No. 1, pp. 140–144. (In Russian)
14. Shahinyan, M.V. To the question of the formation of maximum water discharge of the rivers of the Armenian SSR. Proceedings of National Academy of Sciences of Armenia. 1963, 16 (6): 47–56. (In Russian).
15. Shutov A.M. Calculation of the peak flood discharge on small rivers of the Republic of Mordovia // Bulletin of the Mordovian University, 2003. No. 3-4, pp. 106-112. (In Russian)
16. Surface-Water Resources of the USSR. V. 9. Issue. 2. Moscow: Gidrometeoizdat, 1973: 472 p. (In Russian).
17. Zhang X., Feng, B., Zhang, J., Xu Y., Li J., Niu W., Yang Y. The detection of flood characteristics alteration induced by the Danjiangkou reservoir at Han river, China. Water 2021, 13, 496. https://doi.org/10.3390/w13040496

Methods for long-term forecasting of water availability in spring floods (r. Arpa – p. Jermuk)

Varduhi Margaryan [1,*], *Ekaterina* Gaidukova[2] , and *Gennady* Tsibulskii[3]

[1]Yerevan State University, Department of Physical Geography and Hydrometeorology, Faculty of Geography and Geology, Yerevan, Armenia
[2]Russian State Hydrometeorological University, Institute of Hydrology and Oceanology, Department of Engineering Hydrology, St. Petersburg, Russia
[3]Siberian Federal University, Institute of Space and Information Technologies, Chair of Systems of Artificial Intelligence, Krasnoyarsk, Russia

Abstract. The article discusses the main physical and geographical factors, affecting the runoff of spring floods in the Arpa rivers catchment in the station Jermuk. Also the article discusses the development of a methodology for long-term forecasting of runoff volume of spring flood (WIV–VI) of river Arpa, station Jermuk. The study used data of water discharges of Arpa river catchment (station Jermuk), air temperature, precipitation, reserve water in snow at meteostation Jermuk. A linear correlation was also revealed between the values of the annual runoff and runoff of spring floods in Arpa river catchment, which can be used to predict the annual runoff. To predict the volume of spring flood runoff, regression method and obtained multivariate correlation dependencies. Assessment of statistical significance and stability the proposed models showed their «satisfactory» quality and the possibility of using in the practice of engineering and hydrological forecasts.

1 Introduction

Flood water is considered extreme and becomes a flood, if maximum water flows are formed and, accordingly, high water levels are formed, at which low-lying parts of settlements, agricultural land, roads and railways are flooded, and industrial facilities are damaged [8]. Due to global climate change on the Earth, the number of catastrophic floods has increased, causing enormous damage. But floods cause both damage and great benefits, because due to flood reservoirs, electricity is generated, crops grow, and water is supplied to the population and industrial facilities [1].

Therefore, development of a methodology for long-term forecasting of flood water availability is of significant interest for water consumers and is an important task, solution thereof can help prevent economic damage and human losses, as well as regulate and manage water resources. Taking into account long-term forecasts of river flow, the most important operational decisions are made [2] related to operation of water management systems and implementation of flood control measures. Long-term hydrological forecasts are widely used

* Corresponding author: vmargaryan@ysu.am

to service many sectors of the country's economy, mainly water power engineering, agriculture and water management, timber rafting, fisheries and utilities, water, rail and road transport.

The main objective of this paper is to determine and analyze hydroclimatic conditions for formation of floods in the upper reaches of the Arpa River, and develop a methodology for long-term forecasting of flood water content (runoff volume, W_{IV-VI}) of the Arpa River – Jermuk Village.

The station of the Arpa River – Jermuk Village was chosen as a baseline for the analysis of hydroclimatic conditions of formation flood and forecasting method. It was caused by the following: a) the runoff observation period is 63 years - from 1957 to 2019; b) this station is located in the upper reaches of the Arpa River, above the Kechut Reservoir, at an altitude of 2034 m; d) exclude the influence of the anthropogenic factor as much as possible.

2 Study areas, Data and Methods

To solve the set tasks, the corresponding studies have been used as theoretical basis [3, 5–6]. Daily factual data of "Center for Hydrometeorology and Monitoring" SNCO of the Ministry of Environment of the Republic of Armenia for a period of 60 years and more were used as a source material in the paper. To analyze hydroclimatic conditions for formation of the flood runoff volume and develop a forecasting methodology, the daily water discharge in the Jermuk Station of the Arpa River, the average daily air temperatures and daily precipitation at the Jermuk meteorological station for the period from 1957 to 2019, the maximum water reserve in the snow cover at the Jermuk Weather station were used.

Forecasting methods are developed mainly on the basis of empirical dependencies arising from certain physical considerations based on observational data for the past and expressed in the form of correlative dependences of runoff on the main components of the water balance. In these dependencies, the water reserve in the snow cover and liquid precipitation are the main variables (components). As an indicator of pool moisture, losses for absorption into the soil and evaporation, their indirect characteristic – air temperature of a particular period is sometimes used [10].

When choosing one or another calculation formula, a mathematical model of runoff or flow-forming factors, one should pay special attention to the appropriateness of its use, because for the practice of forecasting and assessing the risk of hazardous spills on rivers and inland water bodies, it is its practical applicability and required reliability of the results, rather than physical validity or novelty of the approach that is of primary importance [4].

Most of the methods of long-term forecasting of spring runoff currently used in operational practice are based on physical and statistical dependencies and were developed in the 60–80s of the last century [2].

They have used a formalized approach in the paper to develop a method for forecasting flow volume of spring floods (from April to June) – obtaining a mathematical dependence that allows calculating future value of the process, namely, statistical multiple regression model, when dependence of the future value on the past is given in the form of an equation. The chosen approach for forecasting hydrological characteristics of river runoff has both a number of advantages and certain disadvantages. In this case, the disadvantage refers to extreme limitation of the available hydrometeorological information. Multivariate statistical model for forecasting the volume of runoff and maximum flow rate of the spring flood of rivers in mountainous areas was used in some papers [1, 8].

Forecasts of spring floods are usually by 25/III, when data are available to assess the hydrometeorological factors responsible for formation of spring runoff. In some anomalous years, forecasts are issued several days earlier or later than the established dates.

3 Results and discussion

Spring flood, summer-autumn flood, summer-autumn and winter low water are the phases characteristic of the water regime of the Arpa River. Spring flood is one of the main phases of the water regime of the Arpa River, which is observed annually in spring or spring-summer. In the station studied, it begins from the third decade of March and first decade of April and ends in the third decade of June and first decade of July, in some cases also in the third decade of July. On average, spring flood water is observed from April 5 to July 7 and lasts 94 days [7]. In the Jermuk section of the Arpa River during the spring flood, 44–69 % of the annual runoff takes place: on average, 57 % (fig.1). Average long-term water discharge at Jermuk Settlement is 5.15 m^3/s. During this period, the absolute maximum costs that can create unfavorable or dangerous situations were recorded for the entire observed period. The maximum flow rates usually pass when rains add up to snowmelt. In Armenia, the period of snow accumulation increases with height and role of snow cover in the flood regime increases.

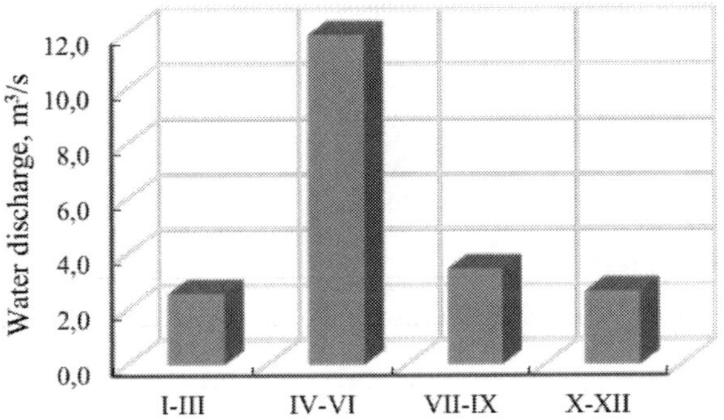

Fig. 1. Intra-annual distribution of water discharge (m^3/s) in the Jermuk Station of the Arpa River

The main factors that determine the power of the spring flood are: water supply in the snow cover before the beginning of snow melting; the amount and intensity of atmospheric precipitation during the period of snow melting and the first phase of flood; autumn-winter soil moistening by the beginning of snow melting, thermal conditions of spring, affecting the intensity of snow melting [8]. With smooth development of synoptic processes, causing a significant increase in air temperature, snow reserves are melting simultaneously in several high-altitude zones and over relatively large areas. In these cases, the flood usually proceeds violently, is characterized by a high rise in the water level and passes over more or less short periods of time, having, as a rule, one wave with a sharply expressed intense rise and a smoother decline. During a protracted spring with a weak development of synoptic processes, the snow cover melts slowly, melt water flows into river beds intermittently, conditions are created for the loss of melt water, which causes a low extended flood with several waves [9].

Due to the diversity of natural conditions of the study area, the volume of runoff and other characteristics of spring floods are distributed very unevenly, but the climate in combination with the features of the relief, soil and vegetation cover, and geological and hydrogeological structure of river catchments play the leading part. Flood characteristics also vary depending on the changing hydroclimatic conditions of formation from year to year. The volume of flood runoff (W_{IV-VI}) ranges from 47.7 million m^3 (1961) to 168 million m^3 (1988), the average value is 94.4 million m^3.

The main factor in the formation of a large volume of spring flood runoff is the amount of water storage in the seasonal snow cover formed throughout the winter, the intensity of snow melting in a friendly spring, the amount of precipitation in April – June in the catchment area, and abundance of liquid precipitation that enhances snow melting. The factor favoring the formation of such values of runoff volumes is the low air temperatures of the winter season, leading to the minimization of losses of the seasonal snow cover (thaw). Among the large volumes of runoff, the floods that took place in 1978, 1987, 1988, 2007, 1988 was the most outstanding. At Jermuk Weather Station, the hydroclimatic conditions were as follows (table 1). In 1988, the maximum values of the total precipitation (570 mm) of the pre-winter and winter periods were recorded, and at this time almost no thaws were observed and the air temperature was - 4.3 °C. Among the small volumes of flood runoff that took place in 1961, 1962, 2015, 2018, the flood of 1961 is the opposite - very low water. In the fall of the previous 1960, the weather was hot and dry; therefore, the moisture content of the basin was significantly below average. The winter was warm, the air temperature exceeded the norm, the amount of snow reserves for the winter of 1960/61 was below the norm.

Table 1. Climatic indicators – factors in the formation of volume runoff flood on the Arpa River at the Jermuk Station

Climatic factors	Years characteristic for water content		
	low water	average	abounding
	1961 г.	1972 г.	1988 г.
Pre-winter and winter precipitation, mm	161	329	570
The amount of precipitation from April to June, mm	185	343	294
Average winter air temperature for pre-winter and winter periods, °C	-3,9	-6,9	-4,3
Air temperature in April, °C	3,7	5,0	3,4

The spring flood looks like a well-defined wave, formed by melted snows, rain and underground waters. Therefore, it is usually of mixed origin. The ascent is intense and is accompanied by individual significant peaks. The duration of the flood rise period is associated with the feeding of the rivers, the passage of the maximum flood discharge. The duration of the flood rise period in the studied section is on average 45 days. The decline in flood as a result of rainfall and groundwater inflow is characterized by the presence of a long plume. And the decline in high water, as in other mountainous regions, in the absence of rain is more prolonged, extended and occurs very slowly and smoothly. The duration of the period of decline in high water compared to the rise is longer and averages 49 days. The decline in high water is mainly determined by the type of feeding, altitude position and the peculiarities of the geological and hydrological conditions of the catchments.

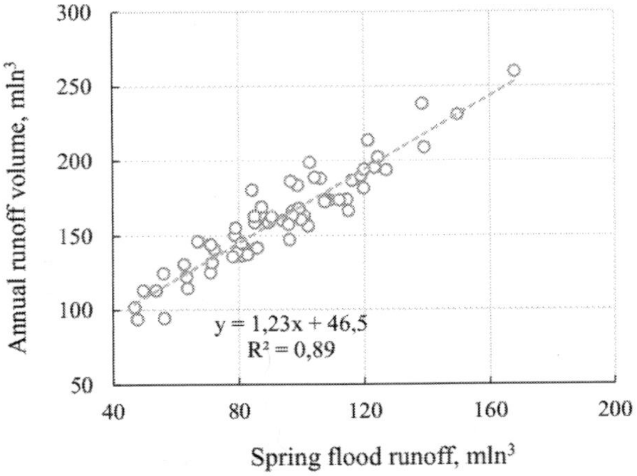

Fig. 2. Correlation between the volumes of spring runoff floods and annual runoff on the Arpa River – Jermuk Station

In the study area, the annual water content of the rivers is closely related to the runoff of the spring flood (fig. 2). As a result of the study, a close correlation was obtained between the actual values of the annual runoff and the runoff of the spring flood on the Arpa River – Jermuk Station. This relationship can be used to forecast annual runoff in the Jermuk Station of the Arpa River.

After researching, analyzing and evaluating the physical and geographical conditions of the basin, available literature sources and initial materials, we obtained dependencies, some of which can be applied when making forecasts.

To forecast the volume of runoff during the flood period, a multifactorial regression method was used, that is, multifactorial correlations between the forecast value and hydrometeorological elements causing it. To obtain these relationships, the volumes of spring flood runoff (W_{IV-VI}) were used as a predictor, and the average monthly discharge in March ($\bar{Q}_{III}$) in the Jermuk Station of the Arpa River, the average air temperature of the Jermuk meteorological station for November-February months ($\bar{t}_{XI-II}$),), for March ($\bar{t}_{III}$), for April ($\bar{t}_{IV}$),), as well as the amount of atmospheric precipitation from December to March ($\sum X_{XII-III}$), from April to May ($\sum X_{IV-V}$), for November ($\sum X_{XI}$). Regression analysis made it possible to obtain the following formulas for calculating the volume of spring flood runoff:

$$W_{IV-VI} = -4{,}67\bar{t}_{XI-II} - 2{,}19\bar{t}_{III} - 2{,}84\bar{t}_{IV} + 3{,}75\bar{Q}_{III} + 0{,}13\sum X_{XII-III} + 0{,}14\sum X_{IV-V} + 0{,}08\sum X_{XI} + 1{,}0 \quad (1)$$

We also tried to use maximum reserves of water in snow (P_{max}) from 1993 to 2019 for the Jermuk weather station as a predictor. In this case, a much better result is obtained. The disadvantage is that the data series for snow stocks is short (1993–2019). The obtained multifactorial correlations for predicting the maximum costs are as follows

$$W_{IV-VI} = -5{,}00\bar{t}_{XI-II} + 0{,}88\bar{t}_{III} - 6{,}35\bar{t}_{IV} + 6{,}30\bar{Q}_{III} + 0{,}24P_{max} + 30{,}5 \quad (2)$$

Indicators of the quality of the forecasting methodology are given in table. 2, where S is the mean square error of the test predictions, σ is the mean square deviation from the norm of the predicted value, R is the correlation coefficient, S/σ is the parameter for evaluating quality of the multiple linear regression model. The development of the method has shown that the obtained dependencies can be used for operational forecasts of the volume of spring floods.

Table 2 Method quality indicators

Formula	Quality indicators		
	S/σ	*R*	Method type
(1)	0,71	0,71	satisfactory
(2)	0,67	0,75	satisfactory

A relationship is noted between the maximum flow rates on snow-fed rivers and the runoff volume during the flood period, and the closest relationship is characteristic of the maximum flow rates with the runoff volume for a pentad or decade, in which the maximum is observed. The closest relationship between the volume and the maximum flow rate of high water takes place on rivers with a predominantly snow supply, with the greatest amplitude of fluctuations in the altitudinal zonality of the catchment area and a relatively short period of snow melting. The least close connection is observed in rivers with mixed feeding. In such cases, the formation of several peaks usually takes place [10]. This pattern is also inherent in the study area. The obtained correlation between the volume (W_{IV-VI}) and the runoff of the maximum flow rates (Q_{max}) for the flood period can be represented by the equation - $W_{IV-VI} = 11{,}2Q_{max} + 39{,}9$. This dependence can be used to predict the volume of flood runoff on the Arpa River – Jermuk Station.

4. Conclusions

Thus, in the Jermuk Station of the Arpa River during the spring flood, 44–69 % of the annual runoff occurs: on average, 57 %.

The volume of the spring flood runoff varies from 47.7 million m^3 (1961) to 168 million m3 (1988), the average value is 94.4 million m^3.

The obtained multivariate dependencies can be used to forecast the volume of spring floods for the Arpa River – Jermuk Station. Individual climatic characteristics of the Jermuk weather station are the main predictors of predictive multiple regression equations.

The proposed methods need improvement by replacing the identified predictors with better ones, with an annual verification of the proposed equations associated with lengthening the predictive series and, accordingly, refining the coefficients.

Financing. This work was supported by the RA Science Committee and Russian Foundation for Basic Research (RF) in the frames of the joint research project SCS 20RF-039 Short-term probabilistic forecast of river flow during the spring flood and RFBR № 20-55-05006\20 accordingly.

References

1. T. Basilashvili, J. Tabatadze, M. Janelidze *Procedia – Social and Behavioral Sciences*. **19** (2011)
2. V.A. Belchikov, S.V. Borsch, Yu.A. Pavroz, A.V. Romanov, M.I. Silnitskaya, Yu.A. Simonov, A.V. Khristoforov *Hydrometeorological research and forecasts*. **4**, 374 (2019)
3. N.N. Karsnogorskaya, Yu.I. Ferapontov, E.V. Nafikova *Proceedings of the Russian State Hydrometeorological University*. **28** (2013)
4. M.V. Kumani, A.V. Apuhtin *Civil Security Technology*. **9**, 4(34) (2012)
5. V.G. Margaryan *Geosphere Research*. **1** (2019)

6. V. Margaryan, E. Fedotova *E3S Web of Conferences*. **149** (2020)
7. V.G. Margaryan, E.V. Gaidukova, L.V. Azizyan, A.E. Misakyan *Water Sector of Russia: Problems, Technologies, Management*. **3** (2021)
8. V.V. Paromov, K.A. Shumilova, I.N. Gordeev *Bulletin of the Tomsk Polytechnic University. Geo Assets Engineering*. **327** (2016)
9. *Surface-Water Resources of the USSR*. **9**, 2 (Moscow: Hydrometeoizdat. 1973)
10. M.V. *Shahinyan The main regularities of formation of the river runoff elements of Armenia and the method of their forecasting* (Leningrad. 1981)

Impact of urban environment on land surface temperature in local climate zones

Anna Gosteva[1*], *Sofia* Ilina[1], and *Aleksandra* Matuzko[2]

[1]SFU, Krasnoyarsk, Russia
[2] Institute of Computational Modelling of the Siberian Branch of the Russian Academy of Sciences (ICM SB RAS), Krasnoyarsk, Russia

Abstract. The replacement of the natural landscape by artificial environment has led to changes in the ecosystem and physical properties of the surface, such as heat storage capacity, and thermal conductivity properties. These changes increase the difficulty of heat transfer between urban areas and the environment. Land surface temperature (LST) images from various satellites are widely used to represent urban thermal environments, which are more convenient and intuitive way. LST maps provide full spatial coverage, which distinguishes them from air temperature data obtained from meteorological stations.
The study of LST according to the Landsat 8 data of Krasnoyarsk city over the past 10 years allowed the authors to talk about the observation of constant seasonal urban heat islands (UHI). For a more detailed consideration of the urban environment, this study further considers urban landscapes, thus the idea of local climate zone (LCZ) is introduced to study these diverse impacts in addition to the traditional map of LST. And analysis of the interaction of UHI and LCZ.

1 Introduction

Humanity has noticed differences in air quality between the city and its surroundings since the 19th century. The city has a different climate, called the urban, which differs from the surrounding area in temperature, wind, wetness and other meteorological variables. Today, these differences are increasing because the climate is changing.

The study is aimed at examining the ecological situation in Krasnoyarsk, studying the perennial heat islands and their relationship with the urban industry [1]. The level of air pollution at constant emission parameters depends on the temperature distribution with height, wind speed and direction, the intensity of solar radiation, air humidity, the amount and duration of precipitation. For low-altitude and cold emissions (chimneys and ventilation exhaust) near the source the concentration of the impurity is low, it increases on the leeward side and reaches a maximum at some distance from the pipe, depending on the wind speed. For high-altitude and hot emissions (chimneys of enterprises of ferrous and non-ferrous metallurgy, some chemical industries, power plants, etc.), the distribution of impurities in the atmosphere depends both on the speed and direction of the wind, and on the vertical

* Corresponding author: agosteva@sfu-kras.ru

distribution of air temperature. With a weak wind, the concentration of impurities near the ground decreases, and with a strong wind, the rate of transport of impurities in the horizontal direction increases.

If the temperature drops with altitude, then in summer during daytime conditions are created for an intense turbulent exchange, which leads to the appearance of significant concentrations of impurities coming from high-altitude sources in the lower surface air layer and to noticeable fluctuations in concentrations with time. If the temperature rises with height in the surface air layer, then the dispersion of impurities reduces. In the case of powerful and long-term inversions at low fugitive emissions, the concentration of impurities can increase significantly [2].

The concentration of impurities in smog can greatly increase due to surface inversion and increased air humidity. Winter smog is often associated with fog, where high concentrations of harmful impurities are retained in the surface air for a long time [3].

With the development of urbanization, the urban heat island (UHI) has attracted a lot of attention due to its adverse effects on the health of residents. UHI was formed from the land surface temperature (LST) of the Landsat 8 data series (https://earthexplorer.usgs.gov/).

Local climate zone (LCZ) is the basis for land cover classification for UHI studies. In this study, we investigated the influence of different seasons UHI in various LCZ of the city. High-resolution series of remote sensing images from OpenStreetMap (https://www.openstreetmap.org/) data was used to map LCZ on a city scale. UHI was formed using Landsat 8 satellite data.

2 Materials and methods

Krasnoyarsk is located at the junction of three geomorphology countries: the West Siberian Plain, the Central Siberian Plateau and the Altai-Sayan mountainous country. Krasnoyarsk is located on both banks of the Yenisei River in its middle reaches. Its width on different banks is not the same: on the right, it varies from one kilometer at the Laletin stream to 8 km at the confluence of the Berezovka River. The left-bank valley expands significantly up to 6 or 8 km [4]. A diversity of landscape and geological structures with water areas and vegetation determines different types of natural landscapes, which determines the development of the city's architecture.

For Krasnoyarsk the wind regime is characteristically constant throughout the year, which is explained by the conditions of orography. Tthe direction of the Yenisei valley coincides with the prevailing wind direction. The width of the Yenisei valley within the city is sharply limited, which affects the speed of the wind flow and, as a consequence, the intensity of ventilation of the territory. Industry is located on the outskirts of Krasnoyarsk [5].

The relief of Krasnoyarsk is distinguished by a variety of forms. Due to the difficult relief of the city, it must be studied with a complex approach.

2.1 The Retrieval of LST

The article describes the application of Landsat 8 OLI/TIRS Collection 2, the dataset Level- 2 includes scene-based global Level-2 surface reflectance and surface temperature science products. Collection 2 includes Landsat Level-1 data for all sensors since 1972, as well as global Level-2 surface reflectance and surface temperature scene-based products from 1982 to present within defined constraints. Landsat Collection 2 Level-2 data became available on the official website only at the beginning of 2021. Access to data of this level removes the question of atmospheric correction of data and determination of atmospheric parameters [6].

Provisional Surface Temperature is generated from the Landsat 8 Collection 1 Level-1 thermal infrared bands, Top of Atmosphere (TOA) Reflectance, Advanced Spaceborne

Thermal Emission and Reflection Radiometer (ASTER) Global Emissivity Database (GED) data, ASTER Normalized Difference Vegetation Index (NDVI) data, and atmospheric profiles of geopotential height, specific humidity, and air temperature extracted from reanalysis data. The quality of the data on the territory of our study is not high enough, the problem became noticeable immediately after the first verification. The Landsat Collection 2 Level-2 data has data loss in the Emissivity layer, and as result in LST layers. Due to partial loss of data, Landsat 8 Collection 2 Level-1 data was used. Figure 1a show results with loss data and figure 1b illustrate the full LST data.

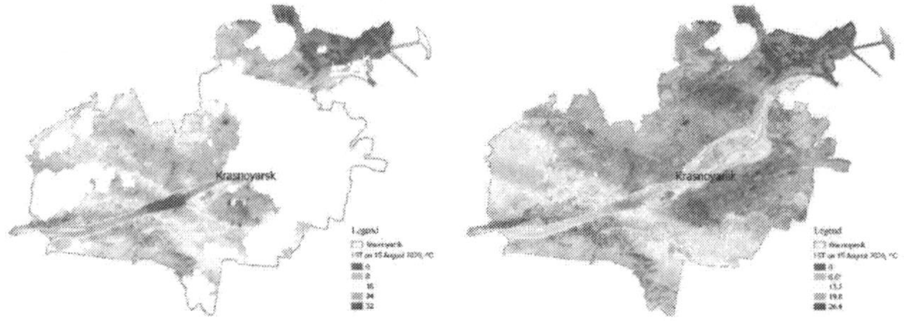

a *b*

Fig.1. The study area represented by LST image retrieved by Landsat 8 on 15 August 2020. (a) The Landsat Collection 2 Level-2; (b) The Landsat Collection 2 Level-1.

The LST for Landsat 8 data have been calculated, the process is described in detail in other works of the authors [7]. QGIS software was used for all calculation. The ground-based observation network of the Federal Research Center of the KSC SB RAS currently has 21 CityAir air monitoring stations. Meteorological stations measure the air temperature 2 meters above the ground, but it is necessary to consider the relationship between air temperature and LST. In addition to the air temperature the monitoring stations give the average value of the concentration in the atmospheric air of PM 2.5 suspended particulate matter.

2.2 Identify the constant UHI

Materials for studying the UHI was collected from the Landsat 8 satellite imagery archive from 2013 to 2020. More than 30 scenes were collected during the snowless period. For each scene, distinguished heat islands were formed using a technique consisting of the following steps:

- Calculate LST from Landsat 8 [7]. Temperature in degrees Celsius.
- For each calculation, determine min, max, average.
- Determine the range of maximum values.
- Convert raster data to vector format.
- Remove the polygons with area less than 2500 m^2.

The perennial UHI was formed in three seasons: spring, summer, and autumn. Since it has been observed that seasonality affects the distribution of the urban heat island.

The accumulated archive of thermal images makes it possible to analyze the dynamics of changes in the UHI from 2013 to 2020. The results obtained were divided into two groups of constant and variable UHI. The group of constants included those heat islands that kept their contours as much as possible from 2013 to 2020. Further, the analysis of the interaction of heat islands and LCZ was carried out.

2.3 Division of urban areas into LCZ

Krasnoyarsk city is being built up. To analyze the dispersion of pollution, in addition to taking into account the features of the relief, it is necessary to pay attention to the development of the city. For this, LCZ were outlined on the territory of the city, which made it possible to reduce various combinations of buildings and land cover to a limited number of classes. LCZ are areas with a uniform surface coverage, structure, materials and a special nature of human activity, and have a specific type of interaction with the surface layer of the atmosphere [8].

This article describes how to use the method of identifying LCZ proposed by Stuart I.D. and Oke T.R. [9] The method is a classification scheme for LCZ, which consists of 17 types of zones, based mainly on the properties of the underlying surface structure of the surface coating, number of storeys of buildings and building density. For Krasnoyarsk the predominant types of development are: the private sector, large-scale buildings and heavy industry, residential areas. It was decided to reclassify the LCZ proposed by Stuart I.D. and Oke T.R. for the relevance of their application for studying the Krasnoyarsk territory.

Also, the methods of microclimatic zoning of Krasnoyarsk's territory are described in the proceedings of K.S. Mokrinets [5], but in his works zoning was carried out according to the geomorphological features of the relief of Krasnoyarsk krai, and in this article microclimatic regions are distinguished by the prevailing type of development.

3 Results

Constant UHI have been identified for three seasons (spring, summer, autumn). Spring heat islands amount to 81 polygons, autumn heat islands amount to 21 polygons, and summer heat islands amount to 82 polygons.

The intersection of autumn UHI and LCZ is located in industrial areas. The autumn UHI are located in industrial areas and residential quarters and in quarters without any buildings. Summer UHI have been distributed across all types of LCZ.

The table shows the correspondence of the air temperature and LST data by 12 meteorological stations of the KSC SB RAS in 2020.

Table. Correspondence of the air temperature and LST.

Observation point	20 May 2020		12 June 2020		15 August 2020	
	Air temperature, °C	LST, °C	Air temperature, °C	LST, °C	Air temperature, °C	LST, °C
Akadem	21,72	31,97	20,15	24,03	20,33	20,93
Shachterov	20,84	36,66	19,45	25,93	20,48	20,16
Partizana	21,18	34,77	20,13	22,7	20,78	20,22
Sverdlovsk	20,42	35,8	21,04	27,45	20,41	18,26
Vetlujanka	19,82	33,35	19,59	26,4	19,58	18,34
Udachni	19,23	34,58	19,56	25,6	20,57	20,43
Pokrovka	20,23	40,15	19,2	20,93	19,97	20,3
Peschanka	21,27	38,06	20,48	27,95	20,82	22,54
Lenina	21,33	37,77	19,92	26,42	20,86	21,97
Sputnik	20,19	36,46	19,71	28,79	19,61	21,38
Kirovsky	20,45	37,58	20,83	27,64	20,17	20,6
KrAZ	23,03	30,92	21,13	24,87	22,08	22,42

The most similar results were obtained from observation on 15 August 2020 (as is shown in Figure 2).

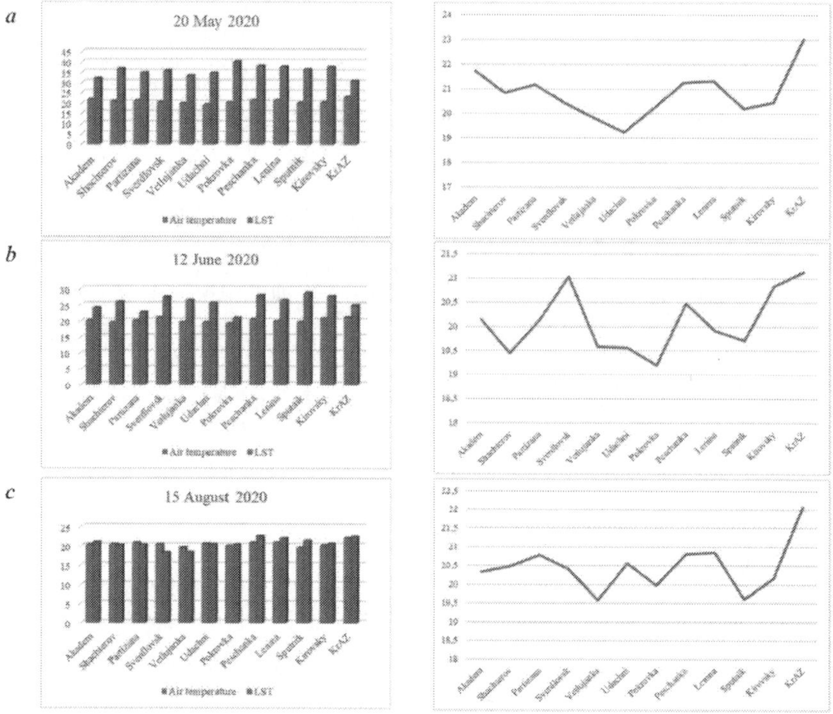

Fig.2. Comparison of the air temperature and LST in observation points. (a) on 20 May 2020; (b) on 12 June 2020; (c) on 15 August 2020.

Having considered the graphs of the obtained temperature values from the monitoring stations, it was found that the values at the observation point "KrAZ" are maximum on all the studied data. It was decided to study the interpolation maps excluding the observation point "KrAZ". Figure 3 a, shows the interpolation by LST on 15 August 2020, this date is chosen as with the greatest coincidence of the data on LST and air temperature. Figure 3b shows interpolated map of air temperature from monitoring stations, and 3c is information about PM 2.5 in the atmospheric air in Krasnoyarsk.

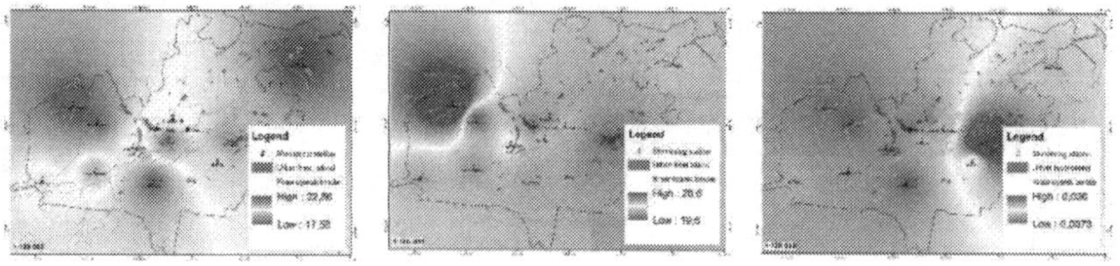

a *b* *c*

Fig.3. Interpolated map on 15 August 2020. (a) by LST image retrieved by Landsat 8; (b) the air temperature from monitoring stations; (c) PM 2.5 from monitoring station.

4 Conclusions

An integrated approach to the study of the urban environment allows us to expand our understanding of the interaction of various factors, for example, industrial facilities, deforestation, difficult landscape, and the state of the environment.

The exclusion of observation points with abnormal temperature rises allows you to see other important parts of the city on interpolation. Taking into account the irregularity of obtaining remote sensing data, usually 3-4 cloudless images per season, it is important to determine constant UHI (preserved for 10 years) and find their relative position with the LCZ. General trends persist, both in the study of LST and the interpolation of air temperature. Studying the readings of pollution, it can be seen that the influence of industrial facilities spreads pollution much further than their territory.

Geospatial and remote sensing tools for urban studies provide an opportunity to plan urban development phases to mitigate the UHI effect. In the following works, socio-economic indicators, such as population density, can be used.

References

1. O. V.Antonenko, V. A. Bezrukikh, E. V.Avdeeva, E. I. Nazarova, A. M. Kislenko Landscape features of the city of Krasnoyarsk as a geological and geomorphological basis for urban planning // Conifers of the boreal area. Vol. XXXV, No. 1-2, P. 15-20 (2017)
2. Grimm, N.B.; Faeth, S.H.; Golubiewski, N.E.; Redman, C.L.; Wu, J.; Bai, X.; Briggs, J.M. Global Change and the Ecology of Cities. S., 319 (2008)
3. Adjustment of the free volume of maximum permissible emissions for the city of Krasnoyarsk [Electronic resource]– http://krasecology.ru/About/PDV (accessed on 24.06.2021).
4. Avdeeva E.V., Wagner E.A., Izvekov A.A. Landscape resources - the basis for the formation of urban greening systems (for example, the city of Krasnoyarsk and its green zone) [Text] // Bulletin of BSU. **2**. 85-90 (2012)
5. Mokrinets K.S. Microclimatic potential of the relief of the territory of the city of Krasnoyarsk as a condition for the formation of the quality of the surface layer of the atmosphere [Text] / K.S. Mokrinets // Bulletin of the Krasnoyarsk State Pedagogical University. V.P. Astafieva. No. 4. P. 295-300 (2011)
6. Landsat 8-9 Calibration Validation Algorithm Description Document [Electronic resource] – https://www.usgs.gov/media/files/landsat-8-9-calibration-validation-algorithm-description-document (accessed on 08.11.2021).
7. Matuzko A.K., Yakubailik O.E., Urban heat island effects over Krasnoyarsk obtained on the basis of Landsat 8 remote sensing data, E. and E. S., **211**, (2018)
8. Samsonov T.E., Trigub K.S., Mapping of local climatic zones of Moscow from space images. G. and C. **6**, 14-25 (2018).
9. Oke, T.R.; Mills, G.; Voogt, J.A. Urban Climates; C. U. P: Cambridge, UK (2017)

TPP ash dump monitoring by ERS methods in the climatic conditions of Siberia

Natalia Kruk[1], *Mikhail* Merko[1], and *Inna* Merko
[1] Cand.Tech.Sc., Siberian Federal University, Krasnoyarsk, Russia

Abstract. A comprehensive monitoring procedure for TPP ash dumps in the climatic conditions of Siberia is proposed, which involves implementation of two stages. At the global stage, remote sensing technology is used directly from spacecraft in Earth's orbit, while at the regional stage, an unmanned aerial vehicle and numerical modeling are used. An example of the operations is given, which clearly shows how exactly this procedure will increase efficiency of the TPP ash dump monitoring in the climatic conditions of Siberia and help promptly take measures required to prevent collapse of the ash dump external soil structure, which may result in ash and slag waste outflow and contamination of the adjacent territory.

At present, the problem of environmentally safe storage of ash and slag from TPPs is very acute. At TPPs of the CIS countries, more than 100 million tons of solid fuel were generated annually from the combustion of solid fuel, and over 1.5 billion tons of ash and slag wastes were accumulated in dumps. TPP ash dumps have a negative impact on the environment (pollution of surface and ground waters during filtration of toxic wastewater, overflow and destruction of dams, wind erosion of stored waste and pollution of the atmosphere and soil, destruction of natural landscapes, and hazardous cryogenic processes). In this regard, the problem of monitoring the increasing TPP ash dump volumes aimed at preventing environmental pollution will become more and more pressing [1].

Local environmental monitoring systems are envisaged to monitor and assess the environmental situation for the operation of ash dumps with earth dams, including an analytical visual and instrumental control complex [2, 3]. Most ash dumps and other similar ponds have local environmental monitoring systems reduced to visual observations, which are carried out through rounds and inspections of structures of this type once every six months, and mandatory recording of the results obtained in the observation log. However, analysis of the observation results and the existing environmental pollution show that this number of visual observations is clearly not enough, for they do not enable prompt identification of the ash dump soil structure deformation at the initial stage. This is especially evident in spring accompanied by a phase transition of the soil massif from frozen to thawed state. This results in the emergence and subsequent development of all kinds of cracks on the ridge, slopes and berms followed by accidents at ash dumps with the ash and slag waste outflow into the adjacent territory, and environmental pollution (Fig. 1).

Fig. 1. Ash and slag waste outflow with contamination of the adjacent territory

Solution to this problem is possible through the use of a dry ash and slag waste storage technique with ash dump formation, which is rather difficult today due to the increased bureaucracy of TPP managing companies and, generally, does not solve the problem of monitoring the state of existing ash dumps. In this situation, increasing the ash dump monitoring efficiency comes to the fore, which relies on remote sensing technology, unmanned aerial vehicles and modern software for processing the observation results obtained. This will make it possible not only to increase the number of control operations, but also to improve the quality of the operation results for the purpose to develop an integrated TPP ash dump monitoring procedure in the climatic conditions of Siberia.

The proposed integrated monitoring procedure should include the following activities:

Global stage. This involves the use of remote sensing technology directly from the spacecraft in the Earth's orbit. The operation must be carried out in accordance with existing regulations once every six months.

As a result, we obtain high-quality images of the enclosing dam outside and the ability to scan the soil massif vertically from the ash and slag deposit upper layer in the pond zone in order to detect initiation of any karst cavities, suffusion outflows and other inside deformation processes, which are not available for visual inspection. Other deliveries may include images of the water level in the pond zone, changes of the settling pond bottom level, stored waste capacity, the speed and power of the sluiced ash and slag material, changes in the structure contours in plan, transverse and longitudinal sections of the enclosing dams, build-up dam body soil, and establishment of soil slopes. The combination of the above data makes it possible to more accurately build a design section of the ash dump necessary for the numerical modeling of industrial wastewater filtration and calculation of the enclosing dam slope stability.

The research data will make it possible to identify and subsequently evaluate:

- changes in the dam crest, slopes, and condition of the structure foot;
- uncontrolled filtration outlets through the dam body and its base;
- points of filtration-suffusion soil removal from the structure and foundation;
- swamping areas in the territory adjacent to the structure foot in the lower pool;
- development of tree and shrub vegetation on the external soil structure within the sanitary protection zone and beyond.

Based on the global stage results, a program of activities is formed to be implemented at the subsequent regional stage.

Regional stage. We envisage the use of an unmanned aerial vehicle for monitoring. This operation can be carried out at the initial stage of information accumulation once a month. Based on the results of processing the information obtained, a more accurate picture of the state of the ash dump and the adjacent territory is formed. This will help more closely monitor the state of the ash dump in accordance with the program of activities developed with the global stage results, in particular:

- all kinds of cracks on the ridge, slopes and berms;
- ice formation on the downstream slope;
- signs of frosty weathering of ash and slag waste;
- development of grass cover and the behavior of earth-moving animals.

This stage will also provide an opportunity for monitoring the state of the environment in the form of observations:

- turbidity in the water of reservoirs formed in the lower pool as a result of filtration;
- erosion and caving of the structure and the banks from the lower pool side.

The program of further activities is then verified following the regional stage results. If any suspicious areas are identified, they are paid special attention, i.e. the number of inspections performed using an unmanned aerial vehicle can be increased up to twice a month, and if an unfavorable scenario develops, these operations can be carried out weekly.

To confirm the relevance of monitoring, let us consider it using the example of ash dump No. 1 at Krasnoyarsk TPP2.

Global stage. The satellite image of ash dump No. 1 is processed in the SAS.Planet.Release.191221 program. The result is shown in Fig. 2a.

Fig. 2a. Results of ash dump No. 1 monitoring at the global stage

Analysis of the image (Fig. 2a) shows that at the time of shooting the ash dump No. 1 was in the following state:

- sections No. 2 and No. 3 are operational, however, at the moment ash and slag is washed into section No. 2, and section No. 3 is already filled and is in the mode of ash and slag dehydration to a moisture content of no more than 50%;
- section No. 1 is intermediate and is used to discharge water from sections No. 2 and No. 3 using bypass pipes located in the separation dams [3];

- section No. 4 is used as a pond for storing clarified water coming from section No. 1 [3].

This means that the main attention should be focused on monitoring the state of section 3, although we continue to monitor the rest of the sections as well.

To obtain more accurate information about the state of ash dump No. 1, we subject the image (Fig. 2a) to analysis using the ENVI 4.4 program. The result is shown in Fig. 2b.

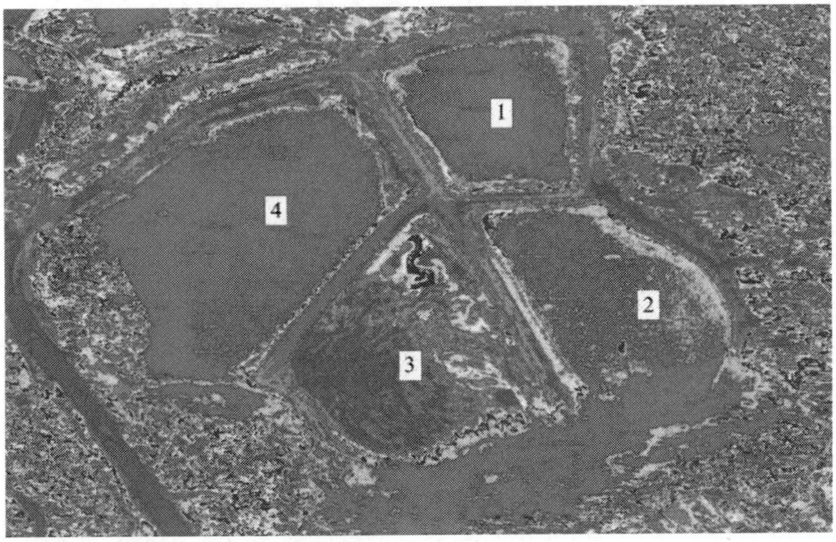

Fig. 2b. Image 2a analyzed using the ENVI4.4.

Analysis of Fig. 2b shows that the main water flow runs to the center of ash dump No. 1, where the system of bypass pipes is located, which means water drainage into section No. 1. However, the right edge of section No. 3 has a clearly visible trail of water flow to the separation road between sections No. 3 and No. 2. This means that this area needs to be taken under more careful control, which we are monitoring at the regional stage.

Regional stage. We collect additional data on the state of the area of special attention and analyze it, which enables us to consider the influence of the data obtained about this part of the ash dump No. 1 on the dependence of the depression curve position on the configuration of the calculated filtration area boundaries. The research is carried out within the framework of filtration analysis in two directions:

- option 1: we take into account the role of the inwashed layer of ash and slag waste when filling section No. 3 of the ash dump No. 1 to a mark of 138 m, and we assume the lower reservoir as the unloading circuit of the filtration flow;

- option 2: we take into account the role of the inwashed ash and slag waste beach of section No. 3 of the ash dump No. 1 from ash and slag waste.

For both options, we perform numerical modeling of filtration in a flat-vertical setting [5]. The algorithm and the calculation program take into account general provisions of the finite difference method and the initial research data. We obtain the filtration flow data, which is further used to form the filtration models for both options of the study as shown in Fig. 3 a and b.

Analysis of the filtration models (Fig. 3 a and b) enables to determine position of the depression curve, distribution of pressure heads in the filtration area, parameters of the seepage area and several types of filtration flow rates.

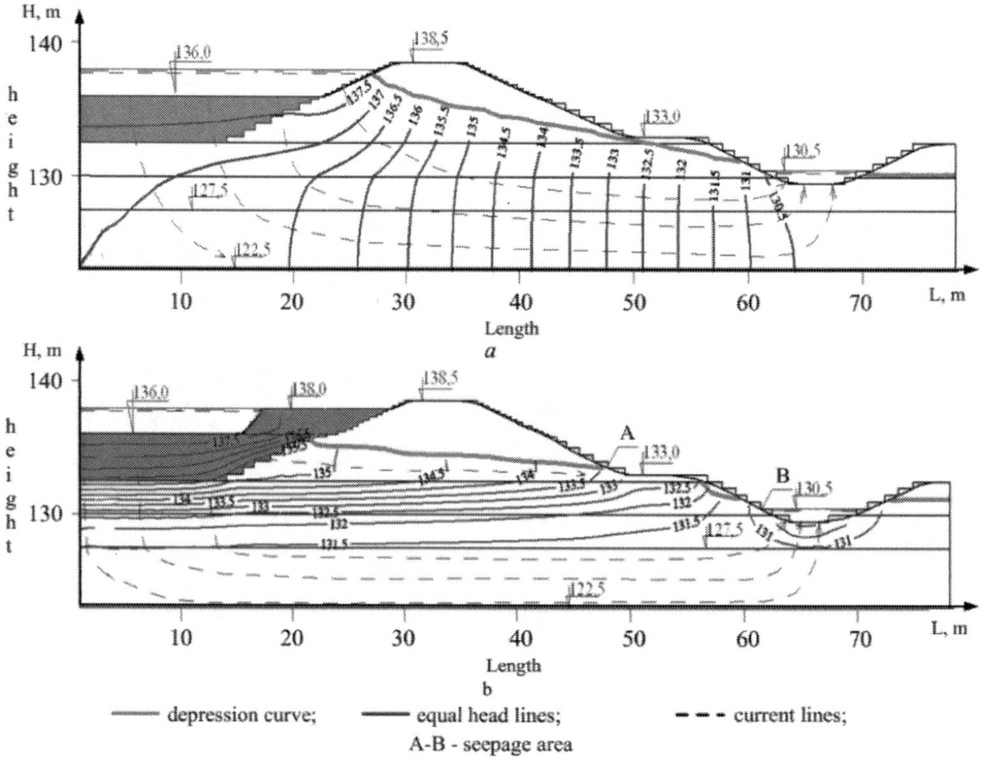

Fig. 3. Model of filtration from the second section of the ash dump: a - Option 1; b - Option 2

Comparative analysis of filtration models between options 1 and 2 (Fig. 3 a and b) shows that the depression curve position depends on the adopted configurations of ash and slag wastes inwash inside the ash dump bowl. In this case, it is clearly seen that in option 2 (Fig. 2b), the depression curve is located much higher than in option 1 (Fig. 2a), which means that the filtration flow seepage runs towards the lower slope and can result in pollution of the environment.

The use of remote sensing methods in monitoring of TPP ash dumps allows for an expansion of the modeling capabilities by using more accurate data due to covering of the following parameters:

- complex configuration of internal and external boundaries of the designed filtration area;
- various boundary conditions, changing in stages during the massif formation;
- the stages of the ash and slag waste massif inwash under the water level in the settling pond.

The proposed integrated monitoring procedure will improve efficiency of monitoring of TPP ash dumps in the climatic conditions of Siberia and help promptly take necessary measures to prevent collapse of the ash dump enclosing dams, which may cause contamination of the adjacent territory as a result of the ash and slag waste outflow.

References

1. G. I. Kuznetsov, N. V. Kruk, Engineering geocryology: study guide. SFU, Krasnoyarsk (2019)

2. GOST R 57793-2017 United power system and isolated power systems. Hydraulic and pumped storage power stations. Hydraulic engineering structures. Monitoring and assessment of technical condition during operation. As of 1 July 2018. Standartinform, Moscow (2017)
3. STO 70238424.27.140.035-2009 Hydroelectric power plants. Monitoring and assessment of the technical condition of hydraulic structures during operation. As of 31 Dec. 2009. Moscow (2009).
4. Draft technical documentation *Ash and slag material obtained as a result of operations of JSC Yenisei TGC (TGC-13) Krasnoyarsk TPP2,* Assessment of the impact of planned activities on the environment, Book 1, Explanatory note, Kemerovo (2017)
5. Y. M. Sysoev, Ash dump design and construction, Energoatomizdat, Moscow (1990)

Image segmentation of acidity of agricultural lands in Eastern Siberia

Dmitriy Perfilyev[1*], *Tatiana* Avdyukova[2] and *Pavel* Zakharov[3]

[1]Siberian Federal University, Institute of Space and Information Technologies, Russian Federation
[2]Director (e-mail: agrohim_24_3@mail.ru)Station of Agrochemical Service 'Solyanskaya', Pervomayskaya str., 19, vil. Novaya Solyanka, Russian Federation
[3]Deputy Head of Library and Publications Department Siberian Federal University, Russian Federation

Abstract. The paper proposes a mathematical model for segmentation of the acidity dynamics of agricultural lands in Eastern Siberia. The segmentation rule is based on the analysis of the image frequency characteristic acidity, which makes it possible to predict the change in the property of the segmented areas. The mathematical approach to the segmentation of the dynamics of agrophysical parameters of agricultural soils has proves to be effective in managing the quality of the natural energy state on the control plots. It is expected that the experience will be applied to the entire responsibility area of 'SAS' 'Solyanskaya' (912.4 thousand ha). Introduction and distribution of innovative practices and technologies will provide new opportunities for monitoring and management of the natural energy state quality of the East-Siberian territory (23 million ha).
Keywords: image segmentation, segmentation rule, segmentation of the dynamics of agrochemical parameters.

Introduction

The national food security problem cannot be solved without a profound digital transformation of technologies for monitoring and predicting the dynamics of key agrochemical parameters of soil productivity. New technologies should be focused on the optimal use and preservation of the natural energy reserve of agricultural lands.

The transformation of technology for monitoring and managing the quality of natural energy reserve of lands requires the widespread use of highly automated (smart) productivity quality management technologies. This should be based on mathematical analysis and prediction of the dynamics of key agrophysical parameters of soils.

Technologically, such transformation can be seen from the mathematical segmentation of the image.

The mathematical formulation of the segmentation problem is discussed in [1-6].

It is shown that the solution to the segmentation problem of the function f (a) is to divide the image into S = {s1, s2, ..., sk}, non-empty connected areas for which the LP rule takes true values only if any pair of elements from any subset of $s_i \in S$ satisfies the goal of interpreter.

Under alternative conditions, the goal is limited to estimating the dynamics of the distribution density of the segmented image properties.

In general, segmentation can be considered as:

* Corresponding author: perfilyev7775@yandex.ru

$$\text{Seg}: f(a) \xrightarrow{Lp} S^{(Sem.)},$$

where:

- $S^{(Sem.)}$ – semantic relations.
- $s_i^{(Sem.)}$ – the name of the $s_i \in S$ *area*;
- $LP(s_i^{Sem})$ – the «neighbourhood» model.

In [20], there are the main forms of syntactic neighbourhood relations that are typical for the image area. For the "Area" model of the image $|s_i|>1$ [20]:

$$s_{10} = < P_i, \{b_i\}^1 \cup \{b_1 – b_3, b_5\}^2 >,$$

where:

- b_i^1 – u-nary attribute of the s_i area that includes a subset of the elements of the P_i. area. For example, it is the sign of the convexity of the shape of the area;
- $b_1^2 – b_3^2, b_5^2$ - models of possible neighbourhood relations of boundary elements of the area. For example, it is the cardinality of the boundary elements of a segmented area with other areas including the «background» area;

A satisfactory segmentation result for S set that is given is an ideal case and is suitable for the "classical" syntax of deterministic mapping of object properties in the image:

- the problem representation allows for setting in advance the value of the iconic function $f(a)$ for the «background» area. The "background" area fills the entire image which contains the S areas representing the properties of the object. The "background" area is defined by the most contrasting value of the function relative to the segmented areas, usually close to the maximum (minimum) value;
- the segmented s_i areas have differences in u-nary features (usually in the gradient: $\nabla f(a)$) and neighbourhood, they are also true for the "background" area. The areas have a boundary in the form of a closed external and possibly internal contour in accordance with the set of neighbourhood models of the boundary elements of the $b_1^2 – b_3^2, b_5^2$ area.

In deterministic conditions, the LP rule is a comprehensive estimation of the $\nabla f(a)$ "distance" (proximity) of the $p_i \in f(a)$ current image element and the neighbourhood (adjacency) of the $p_j \in s_i$ form:

$$|p_i: \nabla f(a) - s_i| \rightarrow min., \text{ and } p_i \cap p_j = 1, \text{ где } p_j \in s_i. \quad (1)$$

Thus, every current element of the $p_i \in f(a)$ image refers only to one $s_i \in S$ area. Accordingly, segmentation is performed on the basis of the uniformity and connectivity of the $f(a)$ image elements.

Without the ideal conditions for displaying the properties of an object, there are regular physical violations of the shape of segmented areas due to distortions in the form of incomplete, false, thickened contours, "overlapping" areas and other related errors [1,10,11].

One of the ways to deal with errors is to modernize the rule (1) of the semantic component, as a configurable "radius" that allows for improving the quality of segmentation. Such a rule has the following form:

$$|p_i: \nabla f(a) - LP(s_i^{Sem})| < r_i^{Sem} \text{ и } p_i \cap p_j = 1, \text{ где } p_j \in s_i, \quad (2)$$

where r_i^{Sem} – the radius of uniformity for s_i^{Sem}.

Thus, the assessment of the uniformity of a class can be changed in the process of segmentation [21, 22], by changing the r_i^{Sem} regarding the analysis of the uniformity of adjacent classes. In addition, an adjustment of the power S can be made on the basis of the analysis of the uniformity of s_i^{Sem} and the power of the "background" area.

The use of the adaptation principle (2) on the one hand improves the quality and on the other hand naturally limits the possibility of applying the $LP(s_i^{Sem})$ rule when using image analysis systems with different properties and in different physical display conditions. Moreover, $LP(s_i^{Sem})$ usually represents a priori knowledge based on iconic information, which limits the practical use of formalization [1-4] in solving the problem of interpretation.

The way out of this situation is to recognize the need to consider the image in a purely mathematical aspect, as continuous and differentiable function $f(z)$, problem-subject area and interpretation of the characteristics of the attributes of segmented image areas [22, 23].

The image of the problem-subject area is based on the given physical attributes of the process (object) under study:

Given $F\{z_1, z_2, \ldots, z_n\}$–the complex of attributes of an object represented by a set of images $f(z_i)$, as well as the rules $Lp(S_i^{Sem})$ that determine the uniformity radius r_i^{Sem} of the i-th attribute value.

The segmentation task can be solved by separating the function $f(z_i)$ into N areas (classes) for which $Lp(S_i^{Sem})$ takes true values that comply with the objectives.

In this case, the segmentation is considered a continuous process of analysing the dynamics of the image $f(z_i)$:

$$\text{Seg}: f(z_i) \xleftrightarrow{Lp(S_i^{Sem})/r_i^{Sem}} n_i^{(Sem.)},$$

where $n_i^{(Sem.)} \in N$ is the semantic relation (class name) defined in the analysis of the uniformity of the current state area $f(z_i)$, limited by the radius $r_i^{(Sem.)}$.

The complexity of segmentation is determined by a number of conceptual aspects:

• interpretation objective –(object phenomenology);

• adequacy of imaging –(of phenomenological properties or essential attributes of the adopted model: $f(z)$).

In the general case, segmentation of the function $f(z)$ satisfies the evaluator when any pair of elements $[z_i–z_{i-1}]$ from any subset $z_i \in f(z)$ compiles the contour description.

$$\oint_C f(z)dz = \lim_{\max[z_i - z_{i-1}]\to 0} \sum_{i=1}^{n} f'(s_i)(z_i - z_{i-1}),$$

where $f'(s_i)$ describes a change in the physical attribute (semantic relationship, area name–S_i^{Sem}); $Lp\ [z_i–z_{i-1}]$ is the reference to the 'neighbourhood' model, defined by the $r_i^{(Sem.)}$ radius of the uniformity of the physical attribute.

The mathematical 'neighbourhood' model for any area of the function $f(z)$ of the closed loop is described by the Cauchy integral formula.

$$f^{(i)}(z) = \frac{i!}{2\pi i} \oint_{CS_i} \frac{f(z)}{(z_0 - z)^{i+1}} dz. \qquad (3)$$

If the path in the positive direction (counter clockwise) is defined by a smooth parametric representation $z = z(t)$ ($\alpha \leq t \leq \beta$), where $z(\alpha) = a$ and $z(\beta) = b$ (a, b are initial arc integration points), then the calculation is carried out according to the formula.

$$f(z)dz = \oint_{\beta\alpha} f(z(t))z'(t)dt = \int_{\alpha}^{\beta} \mathrm{Re}[f(z(t))z'(t)dt + i\int_{\alpha}^{\beta} \mathrm{Im}[f(z(t))z'(t)dt \quad (4)$$

Formula (3) allows for segmenting the contour of area S_i on the image $f(z)$ and (4) describes the contour properties.

In case of a multiple connection (i.e. when bypassing the complement in the negative direction), the segmentation and description of the contour are slightly complicated.

$$f^{(i)}(z) = \frac{i!}{2\pi i} \oint_{CS_i} \frac{f(z_0) - f(z)}{(z_0 - z)^{i+1}} dz \cdot \quad (5)$$

Accordingly, the calculation is carried out by the formula:

$$f(z)dz = \int_{\alpha}^{\beta} \mathrm{Re}[f(z(t))z'(t)dt + i\int_{\alpha}^{\beta} \mathrm{Im}[f(z(t))z'(t)dt + \sum_{s=1}^{n}\int_{S_i} \mathrm{Re}[f(z(t))z'(t)dt + i\int_{S_i} \mathrm{Im}[f(z(t))z'(t)dt \quad (6)$$

It is known that the critical derivation point z_0 serves as the centre of the semantical radius $r_i^{(Sem.)}$. Usually, the value of the critical point is set empirically based on expert knowledge. For example, in the field of plant cultivation, the critical points represent the centres of four groups of cultivated plants with corresponding acidity values.

Conceptually, the value of the critical point is defined on the section of the segmented area boundary and the background where the integral equals to zero at any size and position of the common boundary.

For this reason, it is proposed to connect the position of the critical point with the dynamics of the image frequency characteristic. In this case, the bracketing $r_i^{(Sem.)}$ will ensure the uniformity of the frequency characteristic of the segmented image area.

Method of analysis of the key soil productivity parameters

The important parameters determining the energy state of agricultural lands in the eastern territories of Krasnoyarsk Territory are the following: acidity, moisture, content of humus, active forms of phosphorus and potassium on slightly acidic and neutral soils in the conditions of sharply continental climate of Siberia [16-19].

An example of the segmentation of images of weighted averages of acidity, active phosphorus and uncomplexed potassium for groups of plants on plots of agricultural land in the area of Stepnyaki village, Brazhensky village council, Kansk district, Krasnoyarsk Territory, is shown in Figure 1.

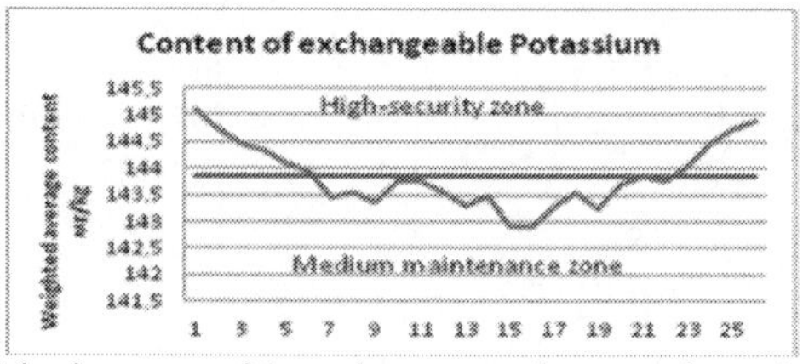

Weighted averages of the exchange potassium, experimental area

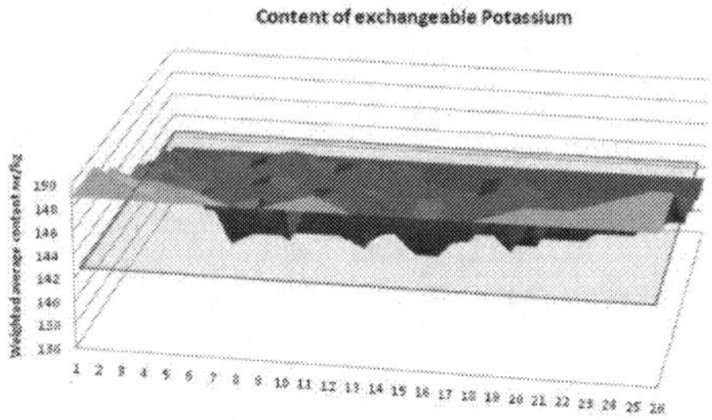

Image segmentation of the distribution of the weighted averages of the exchange potassium

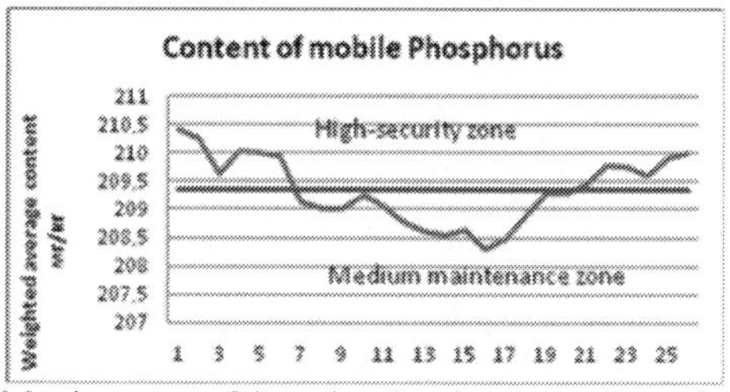

Weighted averages of the active phosphorus, experimental area

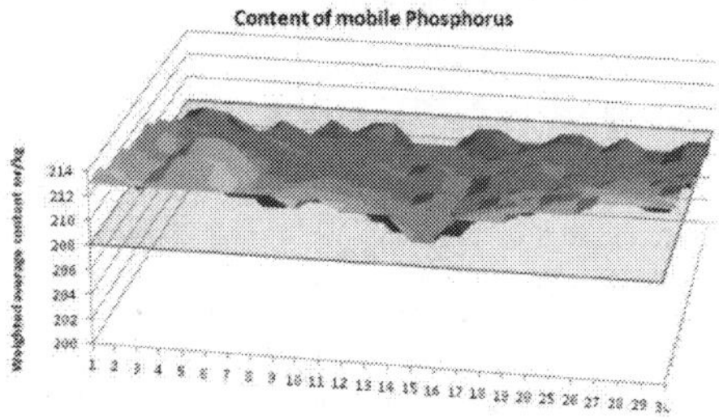

Image segmentation of the distribution of the weighted averages of the active phosphorus

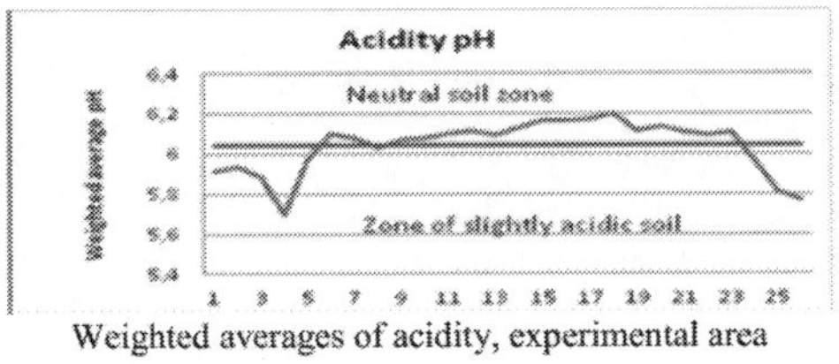

Weighted averages of acidity, experimental area

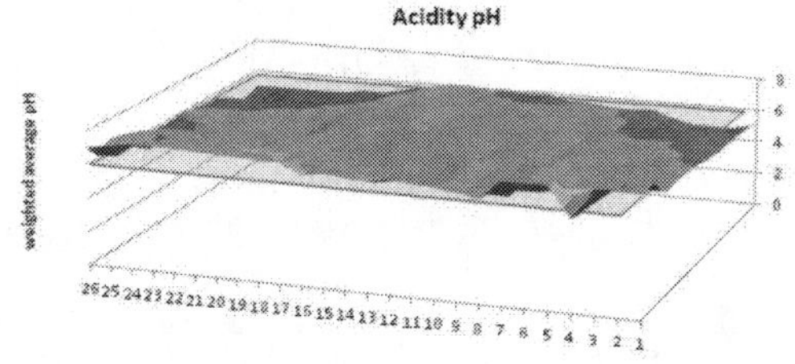

Fig. 1. Image segmentation of the distribution of the weighted averages of acidity

Usually, the interpretation of the object phenomenology is based on the analysis of the frequency, amplitude and phase characteristics of the segmented areas of the image $f(z_i)$.

The segmentation starts from the calculation of the centre line of the image graph $f(z_i)$:

$$\mu_{cp} = \frac{\sum_{i=1}^{k} \mu_i}{k},$$

where k is the number of samples.

The uniformity boundaries are calculated using the Lp rule as follows:

$$Lp_U = \mu_{cp} + \frac{r_i^{(Sem)} \sigma}{\sqrt{n}},$$

where n is the number of measurements.

Deviation value σ is calculated as follows:

$$\sigma = \sqrt{\frac{\sum_{i=1}^{k} \sum_{j=1}^{n} [\mu_{i,j} - \mu_{cp}]^2}{nk}}$$

The segmentation and contour description are carried out in a similar way in case of a multiple connection, i.e. with the integration over the upper and lower boundaries of the image area.

$$Lp_{UL} = \mu_{cp} \pm \frac{r_i^{(Sem)} \sigma}{\sqrt{n}}.$$

Apparently, the segmented areas (see Figure 1) by the L_{pUL} rule should be considered relatively uniform since the segmentation is performed on the basis of a rather abstract parameter like a weighted average.

It is convenient to improve the quality of area segmentation and thus obtain a more accurate solution to the task based on the analysis of their frequency characteristic [20-23]. In this case, the segmentation based on the analysis of the frequency dynamics of the segmented image area $f(\omega_r)$ is described by an integral formula of the type [22,23].

$$f^{(i)}(\omega_0) = \frac{i!}{2\pi i} \oint_{S_i} \frac{f(\omega_r)}{(\omega_r - \omega_0)^{i+1}} d\omega_r \cdot \tag{3a}$$

where: ω_r is the frequency assessment on the interval

$[\omega_i - \omega_{i-1}]$ typical of a slightly acidic medium with a radius of r_i^{Sem}; ω_0 - critical derivation point.

The integral characterising the change in the frequency of the segmented area is calculated using the formula.

$$f(\omega_r)d\omega = \oint_{\beta\alpha} f(\omega_r(t))\omega_r'(t)dt = \int_{\alpha}^{\beta} \mathrm{Re}[f(\omega_r(t))\omega_r'(t)dt + i\int_{\alpha}^{\beta} \mathrm{Im}[f(\omega_r(t))\omega_r'(t)dt \tag{4a}$$

Formula (4a) allows for segmenting the image area with a given frequency characteristic. The analysis of the frequency dynamics in conjunction with the amplitude analysis of dynamics, phase and image area parameters makes it possible to predict changes in the qualitative uniformity of physical parameters under known seasonality conditions and taking into account the quality of the input data.

Presumably, areas with a relatively high frequency index tend to transfer to the more active soil class and vice versa.

Result analysis and discussion

The data on the distribution of the agrophysical parameters on the basis of long-term research conducted by the staff of FSUE SAS 'Solyanskaya' provide a general image of the dynamics of agricultural land productivity in Kansk district of Krasnoyarsk Territory, in particular agricultural lands of Stepnyaki village, Brazhensky village council, Kansk district, Krasnoyarsk Territory.

The calculation of the uniformity radius r_i^{Sem} for the i-th class is carried out from the critical point ω_0 as follows:

$$\omega_{0i} = \mu_{cp} \pm \frac{r_i^{(Sem)}\sigma}{\sqrt{n}} \cdot$$

It should be noted that the value of the 'radius' r_i^{Sem} is described by a graph $f(\omega_r)$. As a result, the characteristic $f(\omega_r)$ makes it possible to form the classes of image areas with the radius r_i^{Sem}, which reflect the uniform object properties (3a).

The calculation of the contour area of the segmented areas is carried out according to

(4a).

Calculation of the contour frequency:

$$f(\omega_r) = P(\omega) + jQ(\omega) = f(\omega)e^{j\varphi(\omega)} = f(\omega)[\cos\varphi(\omega) + j\sin\varphi(\omega)],$$

where: $P(\omega) = \mathrm{Re}\, f(j\omega)$;

$Q(\omega) = \mathrm{Im}\, f(j\omega) = f(\omega)\sin\varphi(\omega)]$;

$\varphi(\omega) = arctg\,\dfrac{P(\omega)}{Q(\omega)}$.

Figure 2, shows the result of the image area segmentation that makes it possible to analyse the dynamics of three acidity classes of soils (slightly acidic, circumneutral and neutral) of the Stepnyaki village agricultural land plot.

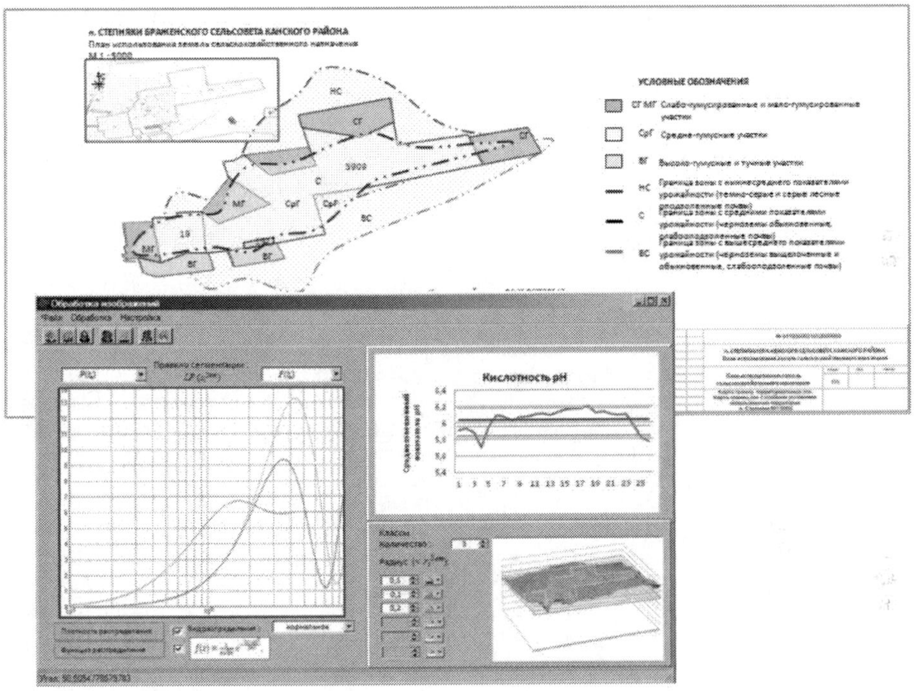

Fig. 2. Segmentation the image frequency properties and evaluation of the uniformity of segmented area classes

The central segmented area covers the entire study area of agricultural land in Stepnyaki village. This area belongs to the 'circumneutral' soil class (pH 5.9), which is relatively to the north of more acidic soils and to the south of neutral soils.

According to further analysis of the central plot in relation to the density of image frequency characteristic $f(\omega_r)$,, the northern part of the plot, with its relatively high frequency parameter, tends to lower the acidity of the medium to a slightly acidic soil (pH 5.6), while the southern part accordingly tends to increase the acidity to a neutral medium (pH 6.2) (as shown in Figure 2. on the right). This is also confirmed by the evaluation of the density of the image frequency characteristic $f(\omega_r)$ of the central plot (Figure 2. on the left).

Overall, the prognosis based on the analysis of the image frequency dynamics shows a relatively high qualitative uniformity of the central plot; the robustness of the plot in maintaining a 'circumneutral' environmental response serves as a good complement to standardised methods of analysis of soil condition of agricultural lands.

Conclusion

The proposed mathematical model is harmonised with the accepted methods used to analyse the main agrochemical parameters [24-28] and greatly enhances the possibilities for more accurate analysis, interpretation and prediction of agrophysical characteristics of soil condition.

The segmentation mathematical model based on more complex 'neighbourhood' relations defined by the uniformity radius r_i^{Sem} of the image area frequency characteristic $f(\omega_r)$ allows for performing the fragmentation of the characteristic into N classes under causal conditions, for which $Lp(S_i^{Sem})$ takes true values that comply with the objectives.

In this case, object segmentation is a combined strategy:

$$\text{Seg: } f(\omega_r) \xleftrightarrow{Lp(S_i^{Sem}) / r_i^{Sem}} n_i^{(Sem.)},$$

where $r_i^{(Sem.)} \in N$ is the semantic relation (object name) determined during the analysis of the frequency uniformity of the segmented area boundaries.

The mathematical approach to the segmentation of the dynamics of agrophysical parameters of agricultural soils has proved to be effective in managing the quality of the natural energy state on the control plots. It is expected that the experience will be applied to the entire responsibility area of 'SAS' 'Solyanskaya' (912.4 thousand ha).

In the future, it is expected to create a 'Digital atlas' reflecting the dynamics of changes in the main agrochemical parameters of soils as part of the Project 'Strategy for the Development of the Agroindustrial Complex of Krasnoyarsk Territory (5 million ha) for the period until 2030 (http://www.krskstate.ru/2030/plan).

Introduction and distribution of innovative practices and technologies will provide new opportunities for monitoring and management of the natural energy state quality of the East-Siberian territory (23 million ha).

References

1. D. A. Denisov Computer aided video analysis methods: monograph. - Krasnoyarsk State Technical University Krasnoyarsk, 1993. - 192 p. (1993)
2. P. A. Chochia Pyramidal algorithm of image segmentation // Information processes. Pub. by Moscow State Technical University named after N. E. Bauman. 2010. **vol. 10, No. 1**.–P. 23–35. (2010)

3. S. Minaee, S., Y. Y. Boykov, F. Porikli, A. J. Plaza, , Kehtarnavaz, N., & Terzopoulos, D. Image segmentation using deep learning: A survey. IEEE Transactions on Pattern Analysis and Machine Intelligence. (2021).

4. M. Brand, V. Kettnaker Discovery and segmentation of activities in video //IEEE Transactions on Pattern Analysis and Machine Intelligence. – 2000. – **T. 22.** – №. 8. – C. 844-851. (2000).

5. S.V. Belim, S.B. Larionov. Segmentation of images on the basis of algorithm of communities separation on the graph // Mathematical structures and modeling 2016. № 3(39). P. 74–85 (2016).

6. G. Khanykov Classification of image segmentation algorithms // The news universities. Instrumentation. 2018. **T. 61**, № 11. P. 978–987. (2018).

7. L. S. Avdyushkina, Two-stage method nature texture classification // Management processes and sustainability 2016 3(1): P. 332-336 (2016).

8. D.N. Gonçalves, V.A. de Moares Weber, J.G. Bragato Pistori, R.d. Costa Gomes, A.V. de Araujo, M.F. Pereira, W.N. Gonçalves, H. Pistori, Carcass image segmentation using CNN-based methods, Information Processing in Agriculture (2020), doi: https://doi.org/10.1016/j.inpa.2020.11.004

9. R. Srikanth and K. Bikshalu, Multilevel thresholding image segmentation based on energy curve with harmony Search Algorithm, Ain Shams Engineering Journal, https://doi.org/10.1016/j.asej.2020.09.003

10. Essam H. Houssein, Bahaa El-dinHelmy, Diego Oliva, Ahmed A.Elngar, Hassan Shaban «A novel Black Widow Optimization algorithm for multilevel thresholding image segmentation» Expert Systems with Applications Available online 4 November 2020, 114159

11. Prakash Purswani, Zuleima T. Karpyn, Khaled Enab, YuanXue, XiaoleiHuang «Evaluation of image segmentation techniques for image-based rock property estimation» Journal of Petroleum Science and Engineering Volume 195, December 2020, 107890

12. TaoWang. ZexuanJi JianYang QuansenSun XiaoboShen ZhenwenRen QiGe «Label Group Diffusion for Image and Image Pair Segmentation» Pattern Recognition Available online 17 December 2020, 107789, 107789

13. Brian L. DeCost, Elizabeth A. Holm A computer vision approach for automated analysis and classification of microstructural image data // Computational Materials Science 110. 2015. P. 126–133. (2015).

14. S. Khodaskar, S. Ladhake Semantic Image Analysis for Intelligent Image Retrieval International Conference on Intelligent // Computing, Communication & Convergence (ICCC-2014) Institute of Management and Technology, India 2014, P. 193–197. (2014).

15. S. Banerjee, S. K. Ghosh, S. Datta, S. K. Saha Segmentation of dual phase steel micrograph: An automated approach // Measurement 2013. P. 2435–2440. (2013).

16. T.V. Avdyukova, S.A. Hrutsky. Fertility State of Arable Lands in the East Zone of Krasnoyarsk Krai // Agriculture. 2017. № 1. C. 25–29. (2017).

17. T. A Milovskih., T. A Vyatkina., T. N. Smirnova Productivity of agricultural land in Sakhalin Oblast conditions // Scientific and technical achievements of AIC, 2019. **Vol. 33. No. 4.** P. 53–56. DOI: 10.24411/0235-2451-2019-10413. (2019).

18. V.V. Chuprova, A.A. Belousov, E.N. Belousova, Yu.V. Gorbunova Evaluation of agroecological condition of the soils involved in the develop-ment of sand and gravel quarries of Kansk district of Krasnoyarsk region // Journal of Siberian Federal University. Agronomic 2019. № 3, 16–21. (2019).

19. Yu. P. Tandelov Productivity of acidic soils in the agricultural area of Krasnoyarsk Territory / Yu. P. Tandelov. - Krasnoyarsk, 2012. - 161 p. (2012).

20. Dmitry A. Perfil`ev Segmentation Object Strategy on Digital Image // Journal of Siberian Federal University. Engineering & Technologies, 2018, **11(2)**, 213-220. (2018).

21. D.A. Perfilyev Object segmentation on a digital image // Regional issues of earth remote sensing; materials of international scient. conf./scient. Rev. by E. A. Vaganov; resp. for rel. A.V. Mashukova - Krasnoyarsk; Sib. Fed. Univ., 2018, 366 p. ISBN 978-5-7638-3109-2. (2018).

22. D.A. Perfilyev Strategy of object segmentation on a digital image // Regional issues of earth remote sensing; materials of VI intern. scient. conf., Krasnoyarsk, September 10 - 13, 2019/scient. rev. by E. A. Vaganov; resp. edit. G. M. Tsibulsky. - Krasnoyarsk; Sib. Fed. Univ., 2019, 368 p. ISBN 978-5-7638-4176-3. (2019).

23. D.A. Perfilyev Segmentation of frequency nonuniformity of the image // Regional issues of earth remote sensing; materials of VII intern. scient. conf. Krasnoyarsk, September 29 - 2 October 2020/scient. rev. by E. A. Vaganov; resp. edit. G. M. Tsibulsky. - Krasnoyarsk; Sib. Fed. Univ., 2020 - 328 p. ISBN 978-5-7638-4389-7. (2020).

24. GOST 26269-91, Soils. General requirements for the fulfilment of analyses Pub.by M.: Standard publishing, (1992).

25. GOST 26204-91, Soils. Determination of mobile compounds of phosphorus and potassium by Chiricov method modified by CINAO. Pub.by M.: Standard publishing, (1992).

26. GOST 26205-91, Soils. Determination of mobile compounds of phosphorus and potassium by Machigin method modified by CINAO M.: Standard publishing, (1992).

27. GOST 26207-91, Soils. Determination of mobile compounds of phosphorus and potassium by Kirsanov method modified by CINAO M.: Standard publishing, (1992).

28. GOST 26483-85, Soils. Preparation of salt extract and determination of its pH by CINAO method M.: Standard publishing, (1985).

Refinement of parameters gravitational anomaly relative to the focal zone of the 2011 earthquake (Japan) based on satellite data

Tatyana Rubleva[1*], *Konstantin* Simonov[2], *Valentin* Kashkin[1], *Anna* Malkanova[1,3], and *Roman* Odintsov[1]

[1]Siberian Federal University, 79 Svobodny pr., 660041, Krasnoyarsk, Russia
[2]Institute of Computational Modeling SB RAS, 50/44 Akademgorodok, 660036, Krasnoyarsk, Russia
[3]Federal Research Centre "KSC SB RAS", 50 Akademgorodok, 660036, Krasnoyarsk, Russia

Abstract. The aim of this work is to study gravitational anomalies that have arisen in the region of the sources of strong underwater earthquakes with a magnitude of $M_w > 8$. For this purpose, data obtained by the GRACE space system were used. Variations of the *EWH* program with a period of 30 days were investigated relative to the focal area of the 2011 Japanese earthquake for the period 2010-2012. It was found that during the preparation of an earthquake, the *EWH* values significantly increase in this area for three months, with aftershock activity, the *EWH* values decrease within a month. Maps of variations of the *EWH* parameter in the conditions of a disturbed geomedia and in background seismic conditions are constructed. The indices of the anomaly δEWH were calculated, which made it possible to analyze in more detail the local gravitational field for the investigated focal zone.

Introduction

In recent years, satellite data has been widely used to analyze geodynamic activity and gravitational changes in areas of increased seismicity. Space systems GRACE (Gravity Recovery and Climate Experiment, NASA (U.S.A.) – GFZ (Germany)) and GRACE-FO (GRACE-Follow-On, of similar design, NASA/GFZ) [1, 2] are successfully used to study the spatio-temporal variations of the Earth's gravitational field. Gravitation satellites of these missions have become an integral part of Earth monitoring.

The technical details of the two space missions can be found in [3]. Global Positioning System (GPS) is used to establish the precise position of GRACE and GRACE-FO satellites. The measurement error on the space system GRACE does not exceed 10 microns, and on GRACE-FO – no more than 1 microns [1].

Note that gravitational fields and their anomalies are studied using the Satellite-to-Satellite Tracking (SST) technique [4]. This technique has been employed in both space systems GRACE (2002-2017) and GRACE-FO (since 2018).

* Corresponding author: tvrubleva@mail.ru

The SST technique consists in the following: the GRACE 1 and GRACE 2 low-orbit satellites are regarded as gravimetric proof mass. Its movement is tracked to obtain the necessary data on changes in the structure of the global (or local) gravitational field over time. Gravitational anomalies observed during seismic events can also be studied using this technique.

The purpose of the present study was to refine parameters of gravitational anomalies that occurred at the epicenters of strong earthquakes of magnitudes $M_w > 8$ that happened in the Pacific Ocean subduction zone, using GRACE data. As a case study, we considered the area of the hypocenter of the catastrophic Japan earthquake of $M_w = 9.1$, which occurred at 05:46:23 UTC on March 11, 2011.

1 Integrative processing of geophysical data

The USGS catalog [5] was used to study geodynamic parameters of seismically active regions of strong underwater earthquakes of magnitude $M_w > 8$: the 11.03.2011 catastrophic Japan earthquake ($M_w = 9.1$), the 24.05.2013 Okhotsk earthquake ($M_w = 8.3$), and three strongest Chili earthquakes: of 27.02.2010 ($M_w = 8.8$), 01.04.2014 ($M_w = 8.2$), and 16.09.2015 ($M_w = 8.3$).

Note that the data of monitoring gravitational anomalies observed during the strongest seismic events in these regions were analyzed using the GRACE measurements [6-8]. Analysis of the spatial structure of the anomalies confirmed the theoretical assumptions about the redistribution of the internal mass of solid Earth during geodynamic activity.

In the present study, based on integrative processing of monitoring data, maps of the epicenter of the 2011 Japan earthquake were analyzed to examine the variations in Equivalent Water Height (*EWH*). This parameter characterizes the equivalent height above the geoid contour [9]. The data on the *EWH* variations above the seismically active region of the catastrophic Japan earthquake with $M_w = 9.1$ were found on website [1].

Software Surfer [10] was used in the stage of integrative processing of the GRACE satellite data. There were missing values in the data obtained. An accepted optimal method of interpolation is Kriging [11]. This method of geostatistics is based on statistical models and autocorrelation analysis. Kriging is fast and enables an accurate fit between the parameter values interpolated into nodes and its initial values. Finally, surfaces of spatial variables are created and their reliability is estimated.

2 Mapping of the *EWH* distribution relative to the earthquake hypocenter

The USGS the size of the focal zone of the study earthquake determined using the MMI scale was 300×100 km [5]. For mapping and further research, we determined the boundaries of the seismically active region of the 2011 Japan earthquake: $\Delta\varphi = 30°$ (25°– 55° N), $\Delta\lambda = 50°$ (120°– 170° E).

The developed procedure was used to visualize the obtained values of the *EWH* variations for the study region. Figure 1 shows the digital map of the *EWH* spatial distribution relative to the region of the perturbed geo-environment of that Pacific Ocean subduction zone, including the coast of Japan. The "perturbed geo-environment" means the seismically active zone with the geodynamic conditions changed due to the seismic processes.

The "cross" in Figure 1 denotes the hypocenter of equation $M_w = 9.1$ with coordinates φ=38.297°N and λ=142.373°E. The hypocenter is 32 km deep. To the right of the map there is an the *EWH* scale (cm).

 https://doi.org/10.1051/e3sconf/202133302011

The anomalous region for the earthquake of $M_w = 9.1$ is oriented longitudinally (South–North) (Fig. 1). The epicenter of the earthquake is between the positive and negative anomalies, which, to a certain extent, coincides with the segment of the boundary between the Okhotsk and Pacific plates.

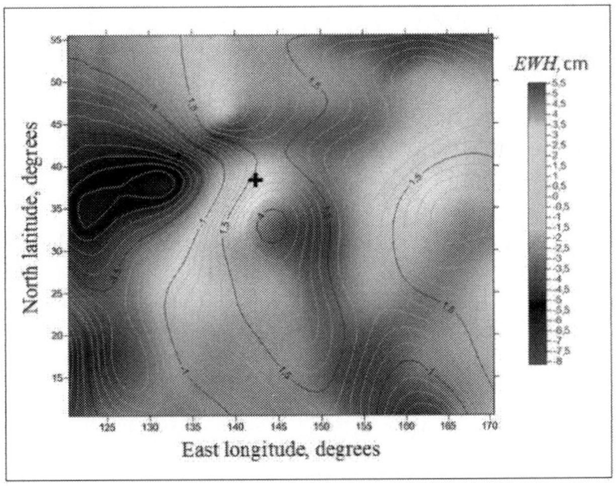

Fig. 1. Spatial distribution of the *EWH* relative to the focal area of the Japanese earthquake

Analysis of the *EWH* spatial distribution suggests that the hypocenter is in the zone of contour line closeness in the northern part of the positive anomalous region. In March 2011, the *EWH* value relative to the epicenter was 1.75 cm. The authors of [6] supposed that formation of the positive anomaly was related to the rising of the bottom of the ocean during the geodynamic activity in the region.

3 Refining the anomaly in the zone of the Japan earthquake based on modified RST

The USGS anomalous changes of gravitational field, to a certain extent, can be associated with geodynamic preparation processes of strong earthquakes. To refine the geometry of gravitational anomalies observed in seismically active regions of strong earthquakes by using the data on the *EWH* variation, we employed RST (Robust Satellite Technique) [12].

Note that this technique was developed for monitoring various natural disasters including earthquakes [13]. In RST, anomaly in the spatiotemporal region is defined as deviation from the unperturbed state of geo-environment. To achieve a more accurate identification, anomaly index is calculated [14].

In the present study, employing the developed procedure, we propose using anomaly index to refine the region characterizing the gravitational anomaly observed in the perturbed geo-environment during seismic activity in the zone of the 2011 Japan earthquake. Taking into account [12], anomaly index was derived from the following formula:

$$\delta EWH(\varphi,\lambda,t) = \frac{\Delta EWH(\varphi,\lambda,t)}{\sigma_{\Delta EWH(\varphi,\lambda,t)}}$$

where φ, λ are geographic coordinates of the image pixel, t is observation date, are values of $\Delta EWH(\varphi,\lambda,t) = EWH_2(\varphi,\lambda,t) - EWH_1(\varphi,\lambda,t)$.

Here $EWH_2(\varphi,\lambda,t)$ are the values of the *EWH* during seismic activity in the perturbed geo-environment, $EWH_1(\varphi,\lambda,t)$ are the values of the *EWH* measured under unperturbed geodynamic conditions, $\sigma_{\Delta EWH(\varphi,\lambda,t)}$ is the standard deviation.

Background conditions (low seismic activity) were revealed by analyzing geo-data using the USGS catalog [5]. In the study region, relatively quiet seismic conditions were observed in January 2008.

Digital maps of anomalies based on the index $\delta EWH(\varphi,\lambda,t)$ were constructed for the epicenter zone of the 2011 Japan earthquake using the developed procedure. Figure 2 shows an example of the anomaly observed relative to the epicenter zone of the Japan earthquake in March 2011. The "cross" denotes the hypocenter of that earthquake with $M_w = 9.1$.

The anomaly with positive values of $\delta EWH(\varphi,\lambda,t)$ relative to the epicenter zone of the Japan earthquake was observed for four months: three months during the earthquake preparation and one month during the aftershock activity. Thus, we revealed spatiotemporal parameters of the *EWH* variation during the geodynamic activity related to the 2011 Japan earthquake.

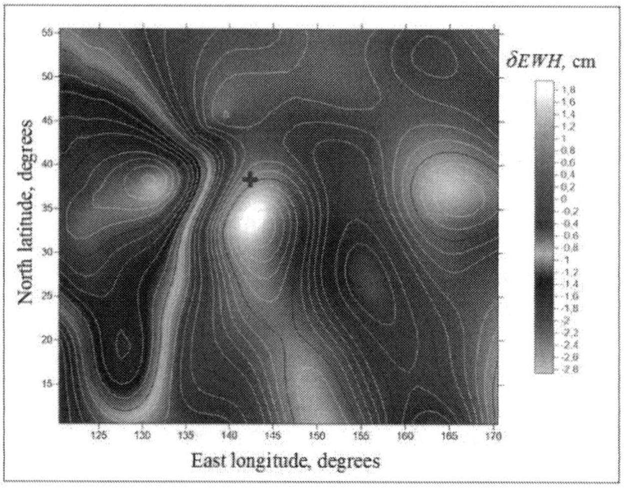

Fig. 2. The area identified by the anomaly index in the area of the 2011 Japanese earthquake

Conclusion

The present study investigated parameters of the focal zone of the Japan earthquake of $M_w = 9.1$ using the 2010 – 2012 data of the USGS catalog. An archive of the data based on measurements of the GRACE space system was compiled. A computational procedure was developed that enabled refining the sizes of gravitational anomalies caused by processes of the strongest underwater earthquakes of $M_w > 8$. Satellite data were used to construct spatial models of the *EWH* variations relative to the focal area of the 2011 Japan earthquake for the perturbed and background conditions of the geo-environment.

We studied thirty-day the *EWH* variations relative to the focal area of the Japan earthquake for the study period. During the earthquake preparation time, for three months, the *EWH* values were found to increase considerably in that zone, decreasing during the aftershock activity. Variations in this parameter relative to the focal area can be caused by various geophysical factors.

The modified RST was used to calculate anomaly indices $\delta EWH(\varphi,\lambda,t)$ and construct refined digital maps of gravitational anomalies relative to the study focal area. The anomaly with positive values of the index was revealed, which characterized the rising of the bottom of the ocean during geodynamic activity in the region.

We think that formation of the anomalous zone was associated with the geodynamic processes in the region. Thus, results obtained in this study suggest that measurements of the GRACE and GRACE-FO gravimetric systems can be used to analyze focal area of the strongest earthquakes.

In our future study, a representative group of the strongest earthquakes in the Pacific region will be used to analyze in detail the *EWH* variations and relationships of this parameter to the global geodynamic and geophysical processes.

References

1. GRACE. URL: https://grace.jpl.nasa.gov/ (Accessed: 14.11.2020)
2. GRACE-FO. URL: https://gracefo.jpl.nasa.gov/news/143/grace-fo-first-science-data-now-available/ (Accessed: 16.11.2020)
3. GFZ/ISDC. URL:https://isdc.gfz-potsdam.de/ (Accessed: 18.11.2020)
4. M.Naeimi, J.Flury, Global Gravity Field Modeling from Satellite-to-Satellite Tracking Data, Springer (2017)
5. USGS. URL: https://www.usgs.gov(Accessed: 7.12.2020)
6. V. Mikhailov, E. Timoshkina, S. Khairetdinov, P. Dmitriev, I. Kartashov, and V.Smirnov, IZV Phys. Solid Earth, **55**, 5 (2019)
7. K. Simonov; V. Kashkin,T. Rubleva,K. Krasnoshekov, E3S Web of Conferences,**75** (2019)
8. W. Qu, Y. Han, Z. Lu, D. An, Q. Zhang, Y. Gao, Remote Sens.,**12** (2020)
9. J. Wahr, M. Molenaar, F. Bryan, J. Geophys. Res. Solid Earth,**103,** B 121 (1998)
10. Surfer. URL: https://www.goldensoftware.com (Accessed: 12.01.2021)
11. N. Cressie, Math. Geol., **22** (1990)
12. V. Tramutoli, V. Cuomo, C. Filizzola, N. Pergola, C. Pietrapertosa, Remote Sens., **96**, 3–4 (2005)
13. A. Eleftheriou, C. Filizzola, N. Genzano, T. Lacava, M. Lisi, R. Paciello, and et al., Pure Appl. Geophys., **173** (2016)
14. F. Marchese, F. Sannazzaro, A. Falconieri, C. Filizzola, N. Pergola, V.Tramutoli, Remote Sens., **9** (2017)

Variations of the total electronic concentration in the ionosphere in seismically active region

Valentin Kashkin[1], *Tatyana* Rubleva[1*], *Konstantin* Simonov[2], *Andrey* Zabrodin[1,3], and *Aleksey* Kabanov[1]

[1]Siberian Federal University, 79 Svobodny pr., 660041, Krasnoyarsk, Russia
[2]Institute of Computational Modeling SB RAS, 50/44 Akademgorodok, 660036, Krasnoyarsk,
[3]Russia[3]Federal Research Centre "KSC SB RAS", 50 Akademgorodok, 660036, Krasnoyarsk, Russia

Abstract. In this work we studied the variations in the total electron concentration (TEC) obtained from measurements of the global navigation system GPS in the preparation zone for the 2010 catastrophic Chilean earthquake (M_w = 8.8) under calm background conditions at a minimum of 24 solar activity (SA) cycles. The analysis of the geodynamic activity and ionospheric TEC disturbances in the seismically active region of this catastrophic earthquake is carried out. A computational technique has been developed that can be used to study TEC variations over seismically active regions.

Introduction

The study of the ionosphere over seismically active regions according to the data of the global navigation systems GPS/GLONASS within the framework of geodynamic monitoring is an urgent task. It is emphasized in [1-3] that movements along a fault in the earth's crust in the region of the earthquake source lead to a significant effect on the atmosphere and emission of gravitational waves along the earth and oceanic surfaces. In [2] it is noted that internal gravitational waves lead to a change in the total electron concentration (TEC) in the ionosphere, which can be recorded by the GLONASS/GPS receiving equipment.

This work is devoted to the study of seismic data in the source zone of the future strong Chilean earthquake of 2010 with a magnitude of M_w = 8.8 and the total electron concentration variations in the ionosphere over these regions under relatively calm background conditions. It should be noted that the preparation of strong earthquakes takes a rather significant time period and requires multidimensional studies over the sources of upcoming seismic events.

The catastrophic earthquake in Chile occurred on February 27, 2010 [4]. The time period of our study is January-March 2009. This period is interesting for two reasons. First, it coincides with the beginning of the 24th solar activity cycle (SA), the formal beginning of which fell on January 2009 [5, 6]. Second, estimates were obtained in [7], from the analysis of which it follows that seismic activity is inversely proportional to solar activity.

* Corresponding author: tvrubleva@mail.ru

The correlation coefficient is R = - 0.8. As a result, in February 2010, Chile [8] experienced one of the most destructive earthquakes in the past 50 years. This earthquake caused a strong tsunami, the height of which reached 3 meters in the coastal zone of the province of Maule (coast of Chile).

The aim of our study is a detailed analysis of seismic data in the source zone of the strong Chilean earthquake of 2010 with M_w = 8.8. At the same time, relatively calm background geophysical conditions were observed. Also, the article refines the total electron concentration variations obtained on the basis of GPS measurements in the preparation area of this natural disaster.

1 Analysis of geomonitoring data

According to the USGS catalog [9] the analysis of geophysical information was carried out in the period from January to March 2009 in the source focal zone of the future Chilean earthquake M_w = 8.8. At the same time, relatively calm background geodynamic conditions were observed. According to updated information from [9] the source size was approximately 600 × 250 km. This zone had an elongated shape parallel to the western coast of Chile in the meridional direction. Its depth according to estimates in [4] reached a value of ≈ 45 km.

In our work according to [9] it was revealed that in January-March 2009, 32 seismic events of weak and medium strength were observed in the studied seismically active region, their magnitudes varied from 3.8 to 5.3.

The distribution of the magnitudes of weak and medium seismic events is shown in Fig. 1. Here, crustal earthquakes occurring at a depth of 70 km are highlighted in blue and intermediate earthquakes with a hypocenter from 70 to 300 km are highlighted in red. The depths of their hypocenters varied from 0.3 to 148 km.

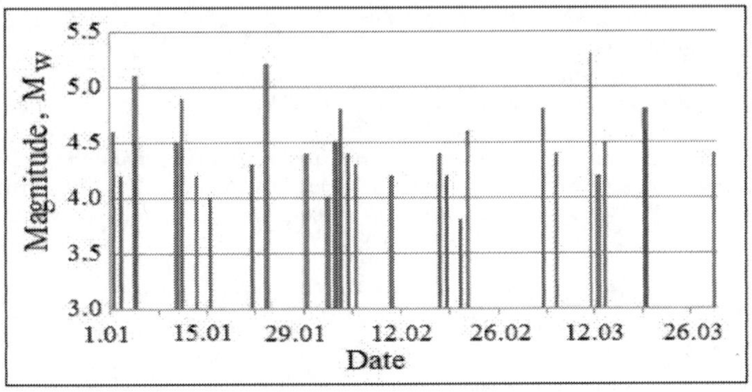

Fig. 1. Distribution of crustal and intermediate earthquakes with M_w > 3 from January to March 2009

The analysis of the spatial location of the studied seismic events in January-March 2009, according to [9] showed that twelve occurred underwater earthquakes were in the ocean near the junction of the two tectonic plates of Nazca and the South American one and the remaining twenty were near the Chilean coast in the transition zone land-ocean.

In the investigated seismically active area as can be seen in Fig. 1 during each month of the 2009 time period there was a sequential change of two dynamic modes in the form of seismic calm and geodynamic activity. The duration of these regimens in time varies from 1

https://doi.org/10.1051/e3sconf/202133302012

day to 2 weeks. Probably, such a rhythm of oscillations in the focal zone of the future Chilean earthquake reflects the alternation of earthquakes differing in magnitude and depth of occurrence. As noted in [10] continuous and different-scale variations in the volume-stress state of the lithosphere at different depths are associated with the processes of interaction of ascending flows of light gases.

The detected oscillatory regime is probably related to the peculiarities of the development of seismicity in the studied region. The results obtained are consistent with the results of work [11] in which the foreshock events of 2000-2010 were considered in the region of the future Chilean earthquake and an increase in the number of earthquakes with a magnitude $M_w > 4$ in 2009 was revealed.

2 Ionospheric precursors of earthquakes

Anomalous disturbances (inhomogeneities) in the ionosphere were repeatedly found above the source zones of strong earthquakes during seismic activity [12, 13]. These atmospheric objects are structural elements of the ionospheric plasma and are recorded in the form of irregular deviations of its electron concentration and other characteristics from the average values [2].

Variations in the total electron concentration are studied as ionospheric precursors of strong earthquakes [12]. According to [14] TEC is the integral number of electrons contained in the vertical column of the atmosphere with a base at the earth's surface and up to the satellite's flight height with a cross section of 1 m^2. The total electron abundance is mainly determined by the electron concentration at the maximum of the ionospheric layer F_2 where the oxygen ion O_+ prevails.

TEC values obtained from GPS measurements are presented in a special IONEX format [15]. We have developed a method for processing this data. An archive of the values of the total electron concentration relative to the focal region of a strong earthquake with $M_w = 8.8$ was formed both for the background conditions in January-March 2009 and for active seismic processes in February 2010.

In Fig. 2 shows the variations in the TEC values from February 13 to March 13, 2010 relative to the focal region of the strong earthquake with $M_w = 8.8$. The TEC unit is: 1TEC = 10^{16} electron/m^2.

This period covers 2 weeks during the period of preparation of the seismic event and 14 days during the period of aftershock activity in the seismically active area. The arrow marks the date of February 27, 2010, when the catastrophic earthquake occurred in the province of Maule (Chile).

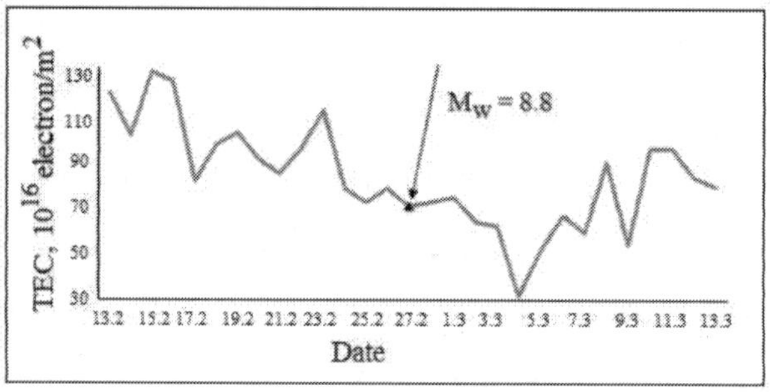

Fig. 2. TEC variations in the ionosphere relative to the epicentral region of the Chilean earthquake

As seen in Fig. 2 within 4 days from 23 to 27 February the total electron concentration values in the ionosphere over the focal area of the strong Chilean earthquake decreased from ($115x10^{16}$ electron/m^2) to ($72x10^{16}$ electron/m^2) by $43x10^{16}$ electron/m^2. In the next 3 days of aftershock activity the values of the total electron concentration over the focal region did not much exceed the value TEC = $72x10^{16}$ electron/m^2 determined for February 27, 2010.

Conclusion

In the course of the research the following results were obtained. A seismic database has been prepared for the source zone of the catastrophic Chilean earthquake of 2010 with $M_w = 8.8$. It was noted that relatively quiet background geodynamic conditions were observed in January-March 2009 at a minimum of 24 solar activity (SA) cycles.

An archive of TEC data obtained on the basis of GPS satellite information for the study periods in 2009 and 2010 has been formed. A method for processing and interpreting TEC data has been developed. A comparative analysis of geophysical and ionospheric data on the source zone of the Chilean earthquake during its preparation in January-March 2009 was performed. A detailed analysis of the total electron concentration values has been carried out which makes it possible to clarify the beginning of the preparation process in the source focal zone of the strong Chilean earthquake.

References

1. V. Kunitsyn, B. Krysanov, A. Vorontsova, Mosc. U. Phys. Bul. Astron., **6** (2015)
2. S. Shalimov, A. Rozhnoy, M. Solovieva, E. Olshanskaya, IZV. PHYS. SOLID EART., **1** (2019)
3. V. Kashkin, R. Odintsov, T. Rubleva, A. Romanov, K. Simonov, *The Lower Atmosphere Response to Seismic Events Using Satellite Data*, In Springer Proceedings in Earth and Environmental Sciences (2019)
4. C. Vigny, A. Socquet, S. Peyrat, J.-C. Ruegg, M. Métois, R. Madariaga, S. Morvan, and et al., SCI., **332** (2011)
5. World Data Center. URL: http://www.wdcb.ru/stp/data.ru.html
6. V. Ishkov, Geomag. and Aeron., **58**, 6 (2018)
7. S. Belov, I. Shestopalov, E. Kharin, DOKL. EARTH SCI., **428**, 1 (2009)
8. M. Farías, D. Comte, S. Roecker, D. Carrizo, M. Pardo, Tecton., **30,** 6, TC6010 (2011)
9. USGS. URL: https://earthquake.usgs.gov/
10. I. Gufeld, O. Novoselov, Bul. KRAUNTS. Ear. Sci., **2**, 34 (2017)
11. B. Derode, R. Madariaga, J. Campos, Sci. Rep., **11**, 2705 (2021)
12. J. Liu, H. Le, Y. Chen, et al., J. Geophys. Res. **116**, A4302 (2011)
13. O. Laryunin, Geodyn. Tectonophys., **10**, 3 (2019)
14. E. Afraimovich, N. Perevalova, GPS monitoring of the Earth's upper atmosphere (2006)
15. IONEX. URL: ftp://cddis.nasa.gov/gnss

Anomalous temperature changes in the UTLS region prior to the 2008 Nura Earthquake

Leonid Sverdlik[1,*]

[1]Research Station of the Russian Academy of Sciences in Bishkek, Kyrgyzstan

Abstract. The paper presents results of retrospective analysis of satellite temperature time series above the epicentral area of the destructive Nura earthquake of M=6.7, occurred in a seismically active Tien-Shan region on October 5, 2008. An algorithm based on the use of a modified STA/LTA criterion has been developed for the purpose of selection and identification of perturbations associated with seismic activity. It has been established that an explicit mesoscale temperature anomaly in the upper troposphere and lower stratosphere (UTLS) was observed during the period from October 1 to 3, 2008. Temporal and spatial distributions of the temperature perturbations consistently appeared at various UTLS levels suggest probable association with seismic event preparation.

1 Introduction

Exploration of thermal anomalies on the Earth surface and in the atmosphere is one of the most promising line of research aimed at establishment of possible connection with seismic activity [1]. Experiments carried out in various geographic regions using satellite data confirmed occurrence of thermal effects before strong earthquakes and their potency to serve as an indicator of tectonic activity [2]. Exploration of processes of interaction between various atmospheric layers and the lithosphere taking place under conditions of strong spatial-temporal variability of the vertical atmosphere temperature structure is both complicated and critical task. It determines the importance of development of effective ways of identification of anomalous variations in temperature changes. In recent decades scientists have achieved a great progress in remote sensing data processing, interpretation of origin atmospheric perturbations origin and development of physical models demonstrating connection between the anomalies and earthquake occurrence [3]. One can mention results of numerous researches in earthquake-prone regions using RST-based (Robust Satellite Techniques) methods [1] and more complex versions of this method to identify pre-seismic anomalies [4].

The algorithm presented in this article based on use of the known STA/LTA (Short Time Average / Long Time Average) criterion that has become a frequent practice of identification of pulsewise oscillations, in particular, recognizing seismic wave arrival and registration of microseismic events [5] may be considered to be one of promising, methods to identify

* Corresponding author: l.sverdlik@mail.ru

anomalies in remote sensing data. Being slightly modified this criterion may be applied to satellite measurements of temperatures.

This study presents an analysis of spatial-temporal temperature variations in the upper troposphere and lower stratosphere (UTLS). Through the example of the Nura destructive earthquake of M=6.7 results of identification of anomalous temperature perturbations associated with seismic activity were demonstrated.

2 Seismic and satellite measurements data

The Nura destructive earthquake of M=6.7 (39.52° N; 73.77° E) occurred in the Southern part of Kyrgyzstan at the depth of ~27.5 km on October 5, 2008 (15:52:49 UTC) [6]. In total throughout the day more than 50 aftershocks of magnitude M=3.0 concentrated near the epicenter of the main earthquake were registered. (Fig. 1a). Concussions covered a vast territory including border areas of Kyrgyzstan, Tajikistan and China. Figure 1b shows the median estimates of peak acceleration as a function of distance [6].

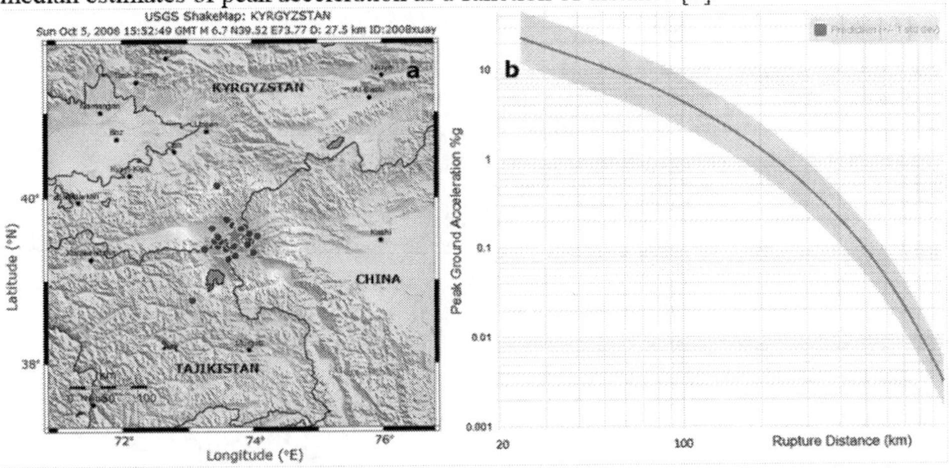

Fig. 1. Aftershock earthquake field M=6.7 (05.10.2008) (a) and distribution of Peak Ground Acceleration in %g (b)

Information of atmosphere temperature changes was obtained from GMAO's Modern-Era Retrospective Analysis for Research and Applications (MERRA-2) [7]. The analyzed set of atmosphere temperature temporal and altitude variations at 12 standard isobaric levels (p_k) from 500 to 40 hPa covered the upper troposphere and lower stratosphere (UTLS). Satellite data presented temperature values in points of uniform grid 0.5°×0.625° and they were restored above the section limited by 35–50° N and 60–90° E for the temporary period from January through December 2008. The temperature data time discretization T(t) was $\Delta t = 3$ h.

3 Satellite measurements processing algorithm

Identification of UTLS temperature perturbations associated with seismic activity was carried out using the algorithm based on analysis of characteristic functions in short (STA) and long (LTA) time windows. On the basis of comparison between seismic activity data and behaviors of temperature time series $T(t_i)$ in UTLS the choice was made in favor of the variance (VAR) which was used as a characteristic function. In this case effectiveness of the algorithm was obviously higher. Calculation of variance ratio (R) was performed for each specific time t_i of temperature data at isobaric levels (p_k):

$$R(t_i, p_k) = \frac{VAR_{STA}(t_i, p_k)}{VAR_{LTA}(t_i, p_k)} \quad (1)$$

Identification of temperature perturbations was based on calculation of anomality value $\delta T(t_i)$ equal to multiplication of two ratios (R_{UT} and R_{LS}) calculated for temperature time series at isobaric levels in the upper troposphere (P_{UT}) and lower stratosphere (P_{LS}). This two-parameter approach allowed to make the algorithm more proof against various non-seismic factors:

$$\delta T(t_i) = R_{UT}(t_i, p_{UT}) \times R_{LS}(t_i, p_{LS}) \geq THR \quad (2)$$

The THR threshold value was established as equal to approximately one standard deviation of the temperature time series being analyzed.

To minimize the probability of forming «false» (in-phase) anomalies parameter δT_C was calculated with account for correlation coefficient $r(t_i)$ between temperature changes at the involved isobaric levels in each short time window (STA):

$$\delta T_C(t_i) = \begin{cases} 0{,}00\_if\_r(t_i)_{STA} \geq 0 \\ \delta T(t_i) \times |r(t_i)_{STA}|\_if\_r(t_i)_{STA} < 0 \end{cases} \quad (3)$$

Values of moving variances in the short window (VAR_{STA}) described local temperature changes occurred during previous 2.5–3.0 days. This value of the window corresponds with the semiperiod range where, where seismo-atmospheric perturbations are assumed to be dominating other sources that determinate temperature changes [5]. Moving variances in the long window (VAR_{LTA}) were calculated according to values of previous 15 days.

Selection of informative layers in the upper troposphere (UT) and lower stratosphere (LS) was made with account for specifics of dynamics of the vertical temperature structure. Examples of vertical profiles of original data (Fig. 2a), of normalized anomalies (Fig. 2b) and variances of temperature fluctuations in the short time window (Fig. 2c) demonstrate that before the Nura earthquake (October 1 through 03, 2008) the most intensive antiphasic temperature changes occurred at isobaric levels 200 hPa (~12 km) and 100 hPa (~17 km). These levels were used to detect pre-seismic perturbations.

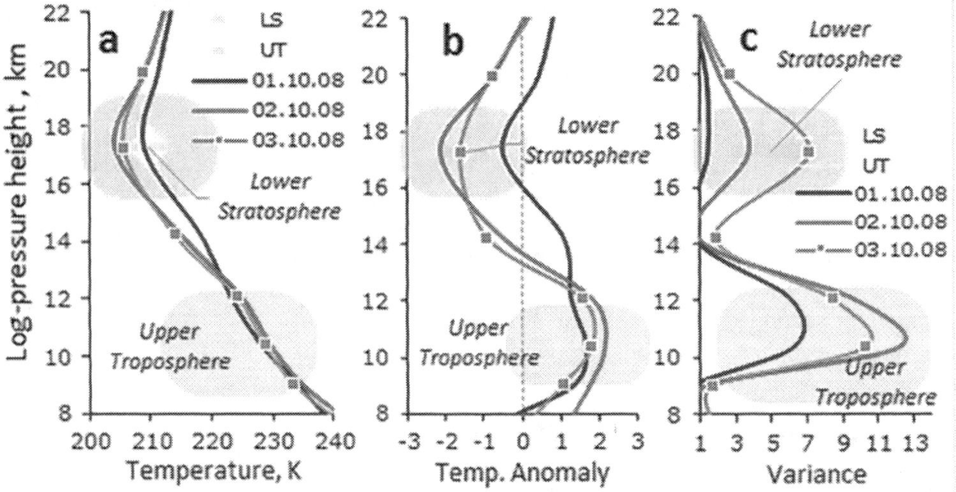

Fig 2. Vertical profiles of temperature (a), normalized anomalies (b) and temperature variances (c) in 8.0–22.0 km altitude range above the epicentral area (38.5–40.5° N; 73.125–75.0° E) of the Nura earthquake of M=6.7 on October 1 through 3, 2008 (00:00 UTC)

4 Consideration of the results

4.1 Identification of perturbations in temperature time series

The altitude-time section of the $R=VAR_{STA}/VAR_{LTA}$ ratio (Fig. 3) demonstrates that before a strong seismic event a consistent change of the UTLS thermal condition due to increase of the amplitude of temperature short-period fluctuations.

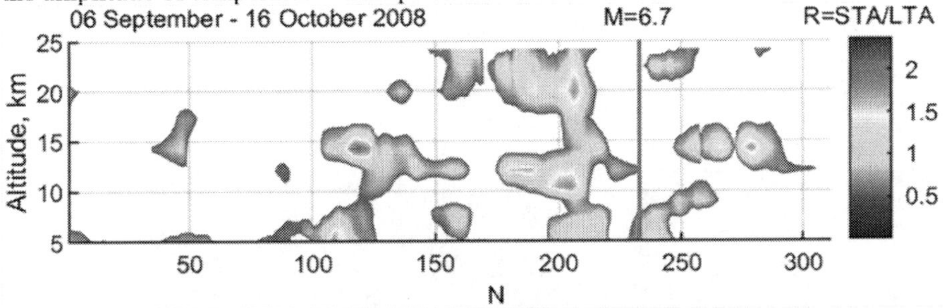

Fig. 3. The altitude-time distribution of the ratio of $R \geq 0.5$ above the epicentral area of the Nura earthquake during the period of September 06 - October 16, 2008

Fig. 4 displays the process of detection of pre-seismic temperature perturbations (Fig. 4a) during the period of September1 - October 31, 2008. If a temperature perturbation is detected the VAR_{STA} variance is changing faster than VAR_{LTA} (Fig. 4b-4c). Accordingly, a substantial increase of the STA/LTA ratio takes place (Fig. 4d). The algorithm maximally uses the coherence effect in anomalous temperature changes, this was clearly manifested on September 5 and October 2–3, 200. The result revealed windows where parameter δT exceed the threshold vale of THR=3.0 (Fig. 4e). As we can see, the most intensive short-period temperature perturbations in the UTLS domain were observed approximately 2–3 days before the earthquake of M=6.7 (Fig. 4f) and they could have been associated with generation of acoustic and gravitational waves [3, 4]. At any other time points of the period under investigation this effect was not observed. Processing of satellite data for a longer period (January-December, 2008) demonstrated that an anomalous temperature change going before the main earthquake stuck out against the background of smaller perturbations [5].

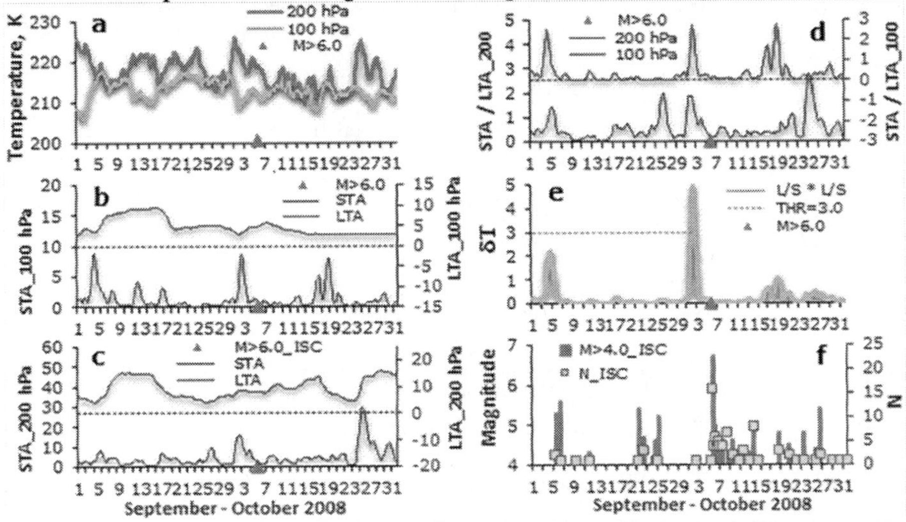

Fig. 4. The initial temperature time series (a); VAR_{STA} and VAR_{LTA} moving variances at levels of 100 hPa (b) and 200 hPa (c); corresponding STA/LTA variances ratio (d) and the product thereof (δT) (e) at the point with coordinates 43.0° N and 73.125° E (markers mark the moment of the earthquake of M=6.7); distribution of magnitudes and the number of seismic events (f) during the period of September 1 through October 31, 2008

4.2 Spatial-temporal distribution of atmospheric temperature anomalies

An important component of pre-seismic perturbation is determination of its spatial localization. On the basis of results of the calculations performed the three-dimensional matrix of dT_C which was used to investigate the dynamics of temperature perturbation distribution in longitudinal and latitudinal directions was restored. The two transversal variability sections of δT_C during the period of September 29 - October 5, 2008 presented in Fig. 5a and Fig. 5b allowed to assess locations and dimensions of anomalous areas, their temporal evolution and to define when δT_C values were maximum.

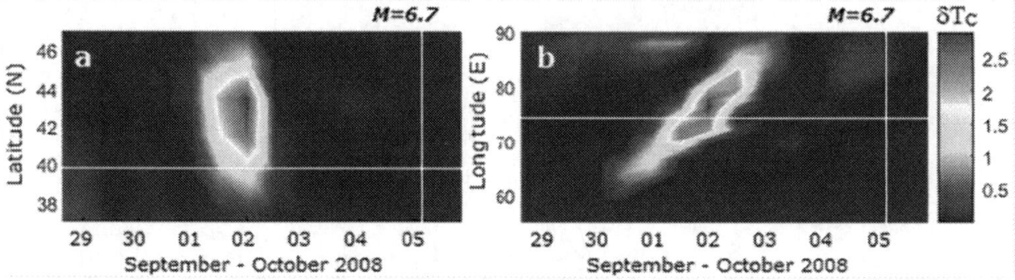

Fig. 5. Distribution of the δT_C parameter by latitude (a) and longitude (b) from September 29 to October 5, 2008. Vertical lines indicate the time of the earthquake (2008-10-05; 15:52:49 UTC). Horizontal lines correspond to coordinates of the earthquake epicenter. Areas where $\delta T_C \geq THR \geq 2.0$ are highlighted in red color

The presented figures demonstrate that UTLS perturbations were observed during the period of October 1-3, 2008. The anomalous area started its formation on October 1 at 09:00 UTC west of the epicenter and then it moved along the latitude of 43° to the east and achieved the maximum on October 2, 2008 three days before the earthquake. The highest values of δT_C=2.79 and δT_C=2.83 exceeding THR=2.0 were observed at 03:00 and 06:00 UTC accordingly. The mesoscale area of abnormally high values of δT_{CORR} was located within coordinates of 40–44° N and 72–80° E (Fig. 6a - Fig. 6d). The maximum of the temperature perturbation observed on October 2, 2008 was displaced by ~2–3° in the north-westerly direction (at 00:00 UTC) and by ~2–4° in the north-easterly direction (at 12:00 UTC) in reference to the earthquake epicenter.

The observed effect of anomaly displacement may be associated with air mass horizontal flows predominant in the atmosphere (~12.0-17.0 km) that were moving at a high speed (above 30 m/s) in the easterly direction (Fig. 6e).

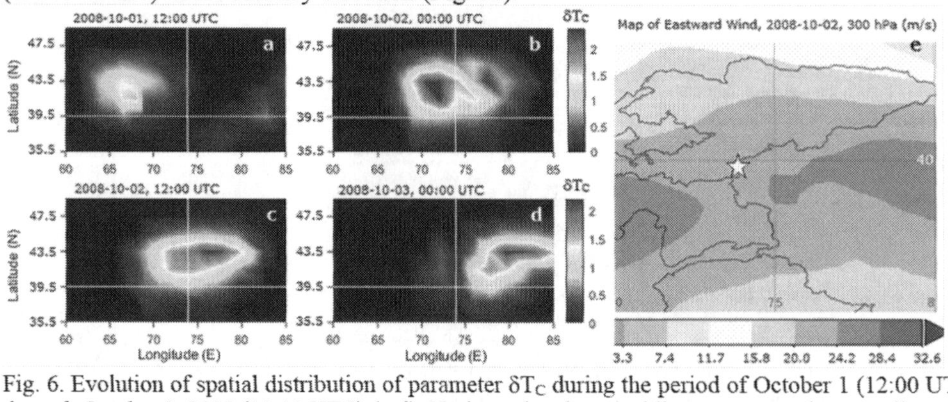

Fig. 6. Evolution of spatial distribution of parameter δT_C during the period of October 1 (12:00 UTC) through October 3, 2008 (00:00 UTC) (a-d). Horizontal and vertical lines correspond to coordinates of the earthquake epicenter of M=6.7 (39.52 °N; 73.77 °E). Map of distribution of eastern component of wind speed at the level of 300 hPa (00:00-23:00 UTC, October 2, 2008) (e)

The quoted results confirm a probable coherence between seismic activity and local temperature perturbation in UTLS that was, steadily observed within 36 hours near the epicenter of the earthquake of M=6.7. On the presumption that acoustic-gravity waves (AGW) [3] are the most probable mechanism of perturbation transfer in the lithosphere-atmosphere-ionosphere system impact of the tropopause inversion layer on vertical distribution of atmospheric waves is obvious. Energy scattering, local heating and turbulence caused by AGW directly or indirectly change the thermal and dynamic structure of the tropopause area [8, 9] making for formation of perturbations in UTLS [10], that become apparent in temperature short-period anomalous variations.

5 Conclusion

Therefore, the presented algorithm based on the use of the modified STA/LTA criterion has sufficient high sensitivity to identify temperature perturbations associated with major earthquakes in the upper troposphere and lower stratosphere. An obvious advantage of the algorithm is simplicity and absence of necessity in additional transformation of initial temperature data.

Effectiveness of the algorithm developed for identification of seismo-atmospheric effects was demonstrated through the example of the Nura earthquake (M=6.7) in Kyrgyzstan. Algorithm testing results reflect the dynamics of temperature perturbations distinctively associated both in time and in space with the seismic process and preparation period of the earthquake under consideration.

Pre-seismic temperature perturbations in UTLS were identified using the developed algorithm and in some other analyzed instances of strong earthquakes (M≥5.0) in the territory of Tien-Shan. However, further study of seismo-atmospheric effects is needed.

Authors enclose gratitude the NASA GES-DISC personnel for providing a free access to satellite measurements data.

The study is partially performed within the framework of the state task of the Federal State Budgetary Institution of Science, Research Station of Russian Academy of Sciences in Bishkek (theme No. AAAA-A19-119020190064-9).

References

1. V. Tramutoli, R. Corrado, C. Filizzola, N. Genzano, M. Lisi, N. Pergola, Bollettino di Geofisica Teorica ed Applicata, **56**, 2, 167 (2015).
2. X. Zhang, C. Kang, W. Ma, J. Ren, Y. Wang, Thermal Science, **22**, 2, 767 (2018).
3. S.-S. Yang, T. Asano, M. Hayakawa, J. Geophys. Res.: Space Physics, **124**, 2, 1410 (2019).
4. L. Sverdlik, S. Imashev, MAUSAM, **71**, 3, 481 (2020).
5. L.G. Sverdlik, Sovremennye problemy distantsionnogo zondirovaniya Zemli iz kosmosa, **18**, 3, 141 (2021).
6. URL: https://earthquake.usgs.gov/earthquakes/
7. URL: https://disc.gsfc.nasa.gov/datasets/M2I3NPASM_5.12.4/summary
8. D. Kunkel, P. Hoor, V. Wirth, Geophys. Res. Lett., **41**, 7822 (2014).
9. Y. Zhang, S. Zhang, C. Huang, K. Huang, Y. Gong, Q. Gan, J. Geophys. Res. Atmos., **120**, 16, 8099 (2015).
10. D. Yu, X. Xu, J. Luo, J. Li, Atmosphere, **10**, 2, 75 (2019).

Study of coherence of temperature variations in the tropopause associated with earthquakes

Leonid Sverdlik[1,*], *Sanjar* Imashev[1]

[1]Research Station of the Russian Academy of Sciences in Bishkek, Kyrgyzstan

Abstract. The paper presents retrospective analysis of satellite data of the upper troposphere / lower stratosphere over the epicentral area of the destructive earthquake of M=6.7, occurred in Tien-Shan territory. We established a well-defined temporal and spatial coherence between temperature perturbations and seismic activity. We used wavelet analysis to identify temporal and spectral coherence of temperature variations in a boundary zone of troposphere and stratosphere before the seismic event.

1 Introduction

Study of pre-seismic atmospheric perturbations performed using ground-based and satellite methods revealed formation of anomalous variations of different geophysical parameters. Thereat, a wide range of physical and chemical processes was identified as short-term precursors of an oncoming strong earthquake [1, 2]. We considered, in particular, releases (emissions) of gases into the atmosphere, anomalous fluctuations of meteorological parameters, electric field intensity, outgoing long-wave radiation (OLR), as well as VLF and ELF bursts, etc. [3, 4].

We used temperature as a parameter changes of which were indications of atmospheric perturbations because it plays a key role in formation of conditions of running of majority of processes and vertical atmosphere structure. While temperature changes are of complicated nature, areas where change of vertical temperature gradient sign takes place are most sensitive to various perturbing factors and atmospheric waves activity [5]. Our investigation is based on analysis of temperature temporal changes at isobaric levels in the upper troposphere and lower stratosphere (UTLS) [6]. Characteristic peculiarities of temperature changes in the atmospheric regions separated by the tropopause allowed investigating temporal and frequency coherence effects preceding major earthquakes [7]

This paper presents study of peculiarities of short-period temperature variations in UTLS over the epicentral area of an earthquake of M=6.7. Dynamic processes in the atmospheric layers and perturbations that are results of interaction between atmospheric layers and lithosphere during periods of seismic activity were analyzed on the basis of remote sensing data.

* Corresponding author: l.sverdlik@mail.ru

2 Seismic and temperature measurements data

The Nura destructive earthquake of M=6.7 was registered in southern Kyrgyzstan (39.53 °N; 73.82 °E) on October 5, 2008 (15:52:50 UTC). The focal depth was ~27 km [8] and about 50 aftershocks with magnitudes varying from 3.0 to 5.9 occurred within 24 hours (Fig. 1a). Fig.1b shows distribution of magnitudes and number of seismic events with M≥4.0 in September-October, 2008.

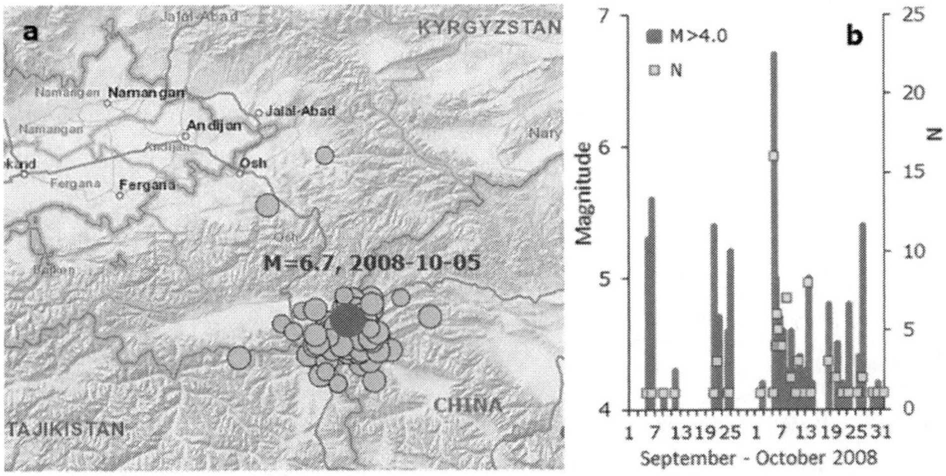

Fig. 1. Aftershock field M=6.7 (05.10.2008) (a) and distribution of magnitudes and number of seismic events (N) with M≥4.0 in September-October, 2008, (b)

Analysis of vertical and temporal variations of atmospheric temperature (T) was performed using satellite measurement data. Temperature data of MERRA-2 global reanalysis system [9] are based on satellite observations processed on a global model and represent synthesized temperature values at standard isobaric levels from 450 to 70 hPa with spatial resolution of 0.5°×0.625°. We considered a period of August 15 – November 15, 2008. The temporal discretization was Δt=3h, which allowed to trace the process of temperature perturbations formation and dynamics of spectrum variation in UTLS in sufficient details.

3 Results and Discussion

3.1 Temporal and spatial distribution of temperature perturbations before Nura earthquake

Fig. 2a shows night temperature profiles (00:00 UTC) in the altitude range of 5.0–25.0 km (450–40 hPa) measured several days before the earthquake in Tien-Shan (October 01–04, 2008.). Initial data of satellite measurements of temperature usually do not allow identifying seismic-atmospheric anomalies and setting them apart from perturbations generated by other natural processes. To investigate interrelation between anomalous temperature changes and seismic activity we can apply methods of satellite data processing, such as RST (*Robust Satellite Techniques*) [2] or modified methods developed from it [6] and based on calculations of normalized temperature anomalies (Fig. 2b) that reflect temperature behavior specifics in the studied region.

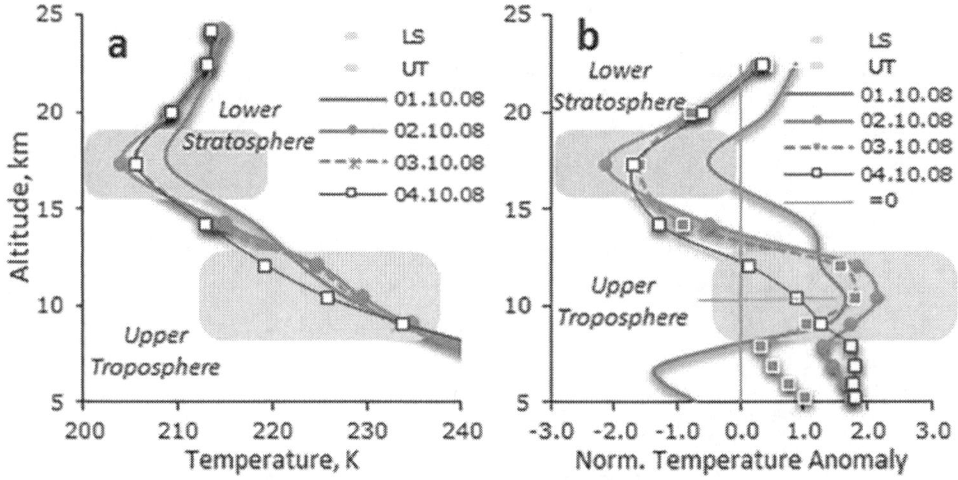

Fig. 2. Vertical temperature (a) and normalized temperature anomalies (b) profiles in the altitude range of 5.0–25.0 km over the epicentral area of the earthquake with M=6.7, October 1-4, 2008 (00:00 UTC)

As it follows from Fig. 2 the most intensive antiphase temperature changes occurred at isobaric levels of 200 hPa (~12 km) in the upper troposphere and 100 hPa (~17 km) in the lower stratosphere. Time series T(t) and normalized temperature anomalies in September – October 2008 corresponding to this levels are presented in Fig. 3a and Fig. 3b accordingly. Results of temperature anomalies diagnosis using the algorithm described in [5, 6] demonstrated that temperature changes in these atmospheric regions occurred in a coherent manner. The well-defined outburst of $D_{\delta T}$ = 2.2, appeared 2 days before the seismic event with M=6.7 (Fig. 3c) indicates a relation with the earthquake preparation period (Fig. 3d).

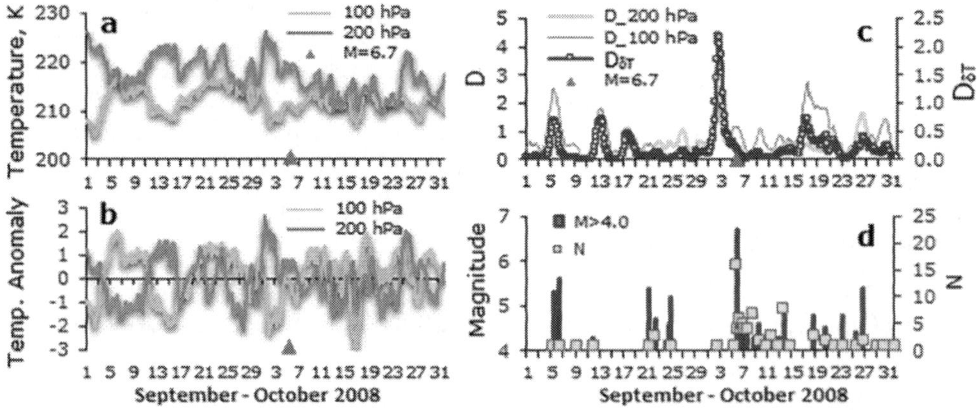

Fig. 3. Temperature (a) and normalized anomalies variations at 200 and 100 hPa levels (b), variations of $D_{\delta T}$ integral parameter (c) and series of magnitudes and number of earthquakes with M≥4.0 (d) in September-October, 2008.

$D_{\delta T}$ integral parameter allowed to assess spatial extent of the temperature perturbation observed in the tropopause during the period of preparation and occurring of the strong earthquake with M=6.7. Results showed formation of a mesoscale temperature perturbation near the earthquake epicentral area. Fig. 4 presents results of processing of temperature data at isobaric levels of 250 and 100 hPa on October 2, 2008 at 15:00 UTC. Horizontal dimensions of the perturbation were several hundred kilometers (in latitude and longitude) and approximately corresponds to the anomaly detected in the course of previous analysis of the seismic event of M=7.4 (December 19, 1992) [7]. High values of $D_{\delta T}$ persisted during 1.5 day and maximum of anomaly was observed on October 2-3, 2008, i.e. ~2.5 days before the seismic event with M=6.7. Spatial configuration of a temperature anomaly is usually substantially affected by atmospheric circulation processes. West-east air transfer in the middle latitudes results in distribution of temperature perturbations in an easterly direction.

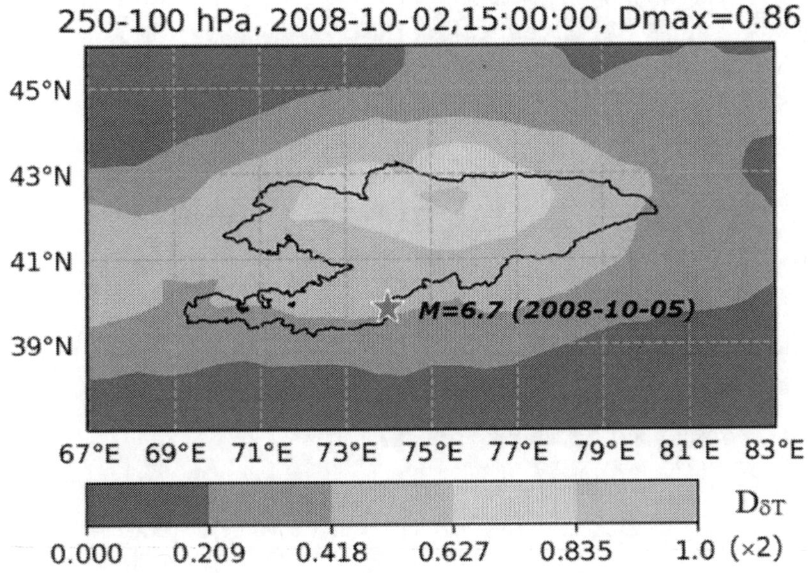

Fig. 4. The spatial distribution of anomalous variation integral parameter $D_{\delta T}$ on October 2, 2008 at 15:00 UTC

3.2 Wavelet Coherency Analysis of temperature variations in the upper troposphere and lower stratosphere

Study of spectral composition dynamics of the data was performed using continuous wavelet transform. Wavelet spectrograms of temperature time series at 200 and 100 hPa levels (Fig. 5) constructed for periods of 0.125 to 32 days demonstrate change of their frequency properties during August 15 – November 15, 2008.

A characteristic feature of the spectrograms is well-defined intensive oscillations within the range of 8–12 days in the beginning of September (100 hPa) and at the end of October (200 hPa), while ~4–5 daily periodicities concurrently appearing at both isobaric levels were registered in the spectrum before the Nura earthquake.

We performed calculation of the spectral coherence measure to analyze effects of consistent variability and statistic interrelation between temperature variations in atmosphere layers separated by the tropopause over the epicentral area of the earthquake. Fig. 6 presents the temporal evolution of the squared coherence between temperature variation spectra at 200 and 100 hPa isobaric levels. The diagramm characterizes the correlations that exist between the spectral components in the temperature oscillations in the upper troposphere and lower stratosphere.

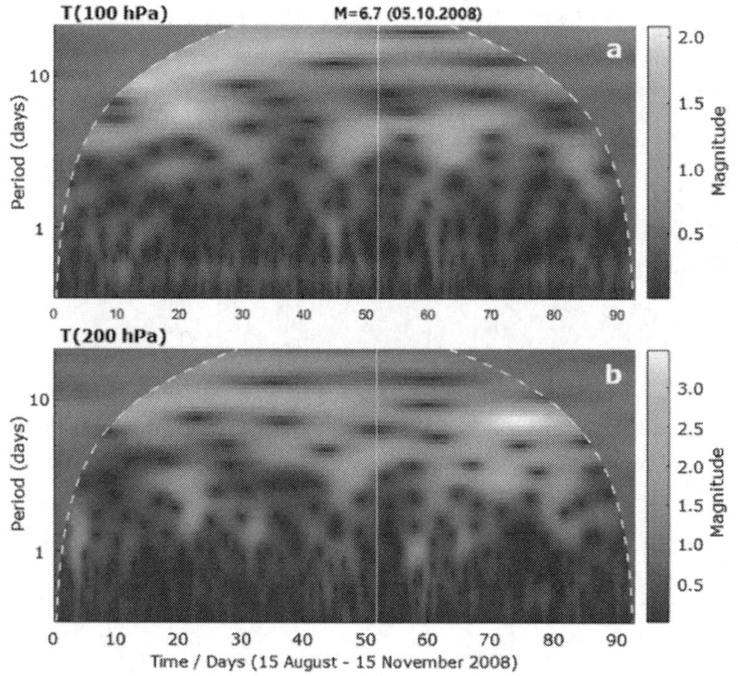

Fig. 5. Temperature variations spectra dynamics at 100 hPa and 200 hPa levels during the period of August 15 to November 15, 2008. The vertical line corresponds to the moment of the earthquake with M=6.7

As we can see, time-localized areas of strong coherence appeared at periods of 3–6 days starting on October 1, 2008. Thereat, virtually there was no enlarging of the coherence area towards long or short periods. Accordingly, the scale range with high coherence coefficients exceeding 0.8 remains relatively stable in time until the moment of the earthquake, that is indicative of consistent running of oscillating processes in the upper troposphere and lower stratosphere. Increasing of coherence values stemmed from anomalous short-period temperature changes. Arrows directions indicate presence of phase change between oscillations T(t). Therefore, the identified frequency-temporal anomaly is in agreement with obtained results of temperature anomalies analysis based on, calculations of integral parameters (Fig. 3) [6, 7].

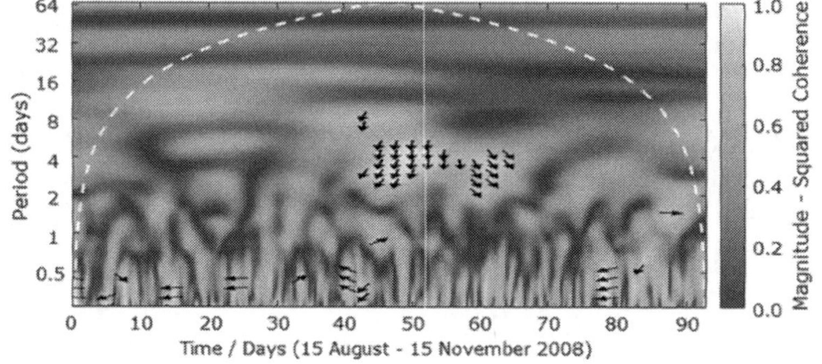

Fig. 6. Evolution of the wavelet coherent coefficients between the temperature variation spectra at isobaric levels of 200 and 100 hPa (August 15 - November 15, 2008). Coherence magnitude and relative phase are denoted by color and orientation of the arrows, respectively

Presented features of temperature coherent are in agreement with results of our previous investigations [7, 10] where we detected coincidence of oscillations frequencies concurrently appearing in temperature variation spectra in the upper troposphere and lower stratosphere. Coherent temperature variations before the Nura earthquake (M=6.7, October 5, 2008) and Suusamyr earthquake (M=7.3, August 19, 1992) can be an indication of pre-seismic perturbations in the tropopause region.

Wave processes are typical mechanism of lithospheric-atmospheric interrelations [3, 11]. A probable reason for formation of anomalous temperature perturbations preceding major earthquakes can be excitement of intense long-period seismic-gravitational oscillations and synchronous and spectrum-identical pulsations that results in UTLS temperature perturbations near the earthquake epicenter [12].

4 Conclusions

Analysis of remote sensing data showed that anomalous temperature changes in the upper troposphere and lower stratosphere were observed over the epicentral area of the strong earthquake with M=6.7. Study of periodicities revealed the effects of coherent temperature behavior in UTLS in the interval of periods varying from 3 to 6 days preceding activation of the seismic regime in Tien-Shan territory.

The presented spatial-temporal distributions of anomalous temperature perturbations point at probable relation to the seismic process.

The obtained results demonstrate prospectivity of application of this method for identification of pre-seismic anomalies in the atmosphere, as well as the necessity of further study of the discovered effect of coherence of temperature variations in the UTLS.

The author is grateful to Giovanni team of NASA GES DISC for free access to satellite measurement data and modeling results.

The study is partially performed within the framework of the state task of the Federal State Budgetary Institution of Science, Research Station of Russian Academy of Sciences in Bishkek (theme No. AAAA-A19-119020190064-9).

References

1. Z-H. Jiao, J. Zhao, X. Shan, Nat. Hazards Earth Syst. Sci. **18**, 1013 (2018).
2. V. Tramutoli, R. Corrado, C. Filizzola, N. Genzano, M. Lisi, N. Pergola, Bollettino di Geofisica Teorica ed Applicata. 56. **2**. 167 (2015).
3. C. Oikonomou, H. Haralambous, S. Pulinets, A. Khadka, S.R. Paudel, V. Barta, B. Muslim, K. Kourtidis, A. Karagioras, S. Inyurt, Geosciences. 11, **16**, 28 (2021).
4. S. Sasmal, S. Chowdhury, S. Kundu, D.Z. Politis, S.M. Potirakis, G. Balasis, M. Hayakawa, S.K. Chakrabarti, Atmosphere, **12**, 1059 (2021).
5. L.G. Sverdlik, S.A. Imashev, J. Sib. Fed. Univ. Eng. Technol. **11**, 956 (2018).
6. L.G. Sverdlik, S.A. Imashev, MAUSAM. **71**, 481 (2020).
7. L. Sverdlik, S. Imashev. E3S Web of Conferences. **223**, 03015, (2020).
8. URL: https://earthquake.usgs.gov/earthquakes/search/
9. URL: https://disc.gsfc.nasa.gov/datasets/
10. L.G. Sverdlik, Sovremennye problemy distantsionnogo zondirovaniya Zemli iz kosmosa. **18**, 3, 141 (2021).
11. V. Kashkin, L. Sverdlik, R. Odintsov, T. Rubleva, K. Simonov, A. Romanov, S. Imashev, E3S Web of Conferences. **149**, 03011 (2020).
12. E.M. Linkov, L.N. Petrova, K.Ts. Osipov, Doklady AN SSSR. **313**, 1095 (1990).

Spatial differentiation and temporal dynamics of drained raised bogs in Western Siberia

Anna Sinyutkina[1*]

[1] Siberian Federal Scientific Centre of Agro-Bio Technologies of the Russian Academy of Sciences, Siberian Research Institute of Agriculture and Peat, Russia

Abstract. This study analysed the influence of drainage on the vegetation cover of raised bogs in the taiga zone of Western Siberia. The study was based on a temporal analysis of Landsat satellite imagery data using the normalised difference vegetation index (NDVI) for the period from 1975–2020. We investigated four key sites within drained raised bogs. The analysis was not carried out using absolute NDVI values; rather, the ratio of the NDVI values of drained bogs to the NDVI values of a similar pristine bog was used. Four stages of the vegetation dynamics of drained bogs were determined. In the first stage, including the period before drainage, the NDVI values were close to those of the pristine site, which confirms that there are similar conditions before drainage. The second stage, from 1989–2001(2002), is characterised by a decrease in NDVI values relative to the pristine bog. This was probably due to the degradation of moss vegetation, which is a sensitive indicator of a decrease in the water table level in the absence of a significant growth of the tree layer. Furthermore, since the 2000s, there has been an increase in the NDVI values and they have stabilised at the level of a pristine bog.

1 Introduction

In the taiga zone of Western Siberia, the drainage of raised bogs was carried out from the 1960s to the1980s, mainly for agricultural and forestry activities, but also for peat extraction. Drained mires occupy more than 500 km^2 (less than 1% of the total wetland area), and among them, pine dwarf-shrub sphagnum raised bogs cover only 130 km^2. However, despite their small area, these sites are at risk from peat fires. About 30% of their total area burned in several different years (1998, 2014, 2016, 2019) [1, 2, https://firms2.modaps.eosdis.nasa.gov/map]. Draining did not result in increased wood biomass, mainly because of poor mineral nutrition. The effect of drainage on vegetation was limited to a change in the species composition, mainly of moss and dwarf shrub cover. Currently, the drained bogs are not used, and the ditch culvert capacity has decreased due to overgrowing. Within the drained mires of the taiga zone of Western Siberia, there are no long-term field studies to assess the impact of drainage and the further development of vegetation cover. Thus, the use of multi-time satellite images is generally accepted, and it is sometimes the only method available for the long-term monitoring of wetlands. In particular,

* Corresponding author: ankalaeva@yandex.ru

many studies have shown that Landsat data can be used to model the characteristics of the vegetation cover of mires; this is an inexpensive and effective method for assessing carbon accumulation in mires of different biogeographic regions [3–9]. Despite its prospects and high scientific and practical significance, for the territory of the taiga zone of Western Siberia, studies on monitoring ecosystem processes using satellite data within wetlands are not numerous [10].

The aim of the study is to evaluate the transformation of vegetation cover after drainage and the intensity of the processes of the self-restoration or degradation of the drained bogs based on a temporal analysis of Landsat satellite imagery data using the normalised difference vegetation index (NDVI).

2 Objects and methods

2.1 Study sites

We investigated five key sites within drained and pristine raised bogs in the south-eastern part of Western Siberia within the Tomsk region. The sites were located within interfluve valleys and terraces in the basin of the Chaya River (Middle Ob basin) and on the left bank terrace of the Ket River (table 1). The studies were conducted within the pine dwarf-shrub *Sphagnum* raised bogs. The most dominant woody species is *Pinus sylvestris*. The tree height ranges from 0.5 to 5 m, with a tree crown projective cover below 10%. The dwarf-shrubs *Ledum palustre*, *Chamaedaphne calyculata*, *Andromeda polifolia* and other species are widespread, with an average cover of 70–90%. In the moss layer, species of the genus *Sphagnum* dominate, including *Sphagnum fuscum*, *S. magellanicum*, *S. balticum* and *S. angustifolium*. Moss cover reaches 90–100% [11]. The peat deposit thickness ranges mainly from 1 to 4 m, sometimes reaching 6 m.

Table 1. Key sites characteristics.

Key site		Coordinates	Geographical position	Altered by human activity
1	Kolpashevo bog (K)	58°19'N 83°07'E	Ket River terrace	Forestry drainage
2	Ust-Bakchar bog (UB)	57°35'N 82°16'E	Bakchar River terrace	Drainage for peat extraction
3	Bakchar bog (BB)	56°54'N 82°41'E	North-eastern part of Great Vasyugan Mire, Bakchar and Iksa Rivers interfluve	Forestry drainage
4	Iksa bog (IB)	56°51"N 83°16'E	North-eastern part of Great Vasyugan Mire, Iksa and Shegarka Rivers interfluve	Forestry drainage
5	Bakchar bog (BBpristine)	56°58'N 82°37'E	North-eastern part of Great Vasyugan Mire, Bakchar and Iksa Rivers interfluve	Pristine bog

Drained pine dwarf-shrub *Sphagnum* raised bogs occupy 130 km^2 within the south-eastern part of Western Siberia. Drainage was carried out in the 1970s and 1980s via a network of open ditches with a distance of 150–180 m on the drained site for forestry (the Bakchar, Iksa and Kolpashevo bogs) and 20–60 m on the drained site for peat extraction (the Ust-Bakchar bog); the planned width was 1–2 m, with a depth of up to 1 m. Draining did not result in increased wood biomass, mainly because of poor mineral nutrition [1]. The lack of planting and repair of the drainage network has led to the development of the self-restoration processes of some drained bogs. Depending on the initial drainage rate and the current state of the drainage network, bogs are at different stages of the self-restoration or degradation of vegetation cover. A key site in the Klyuch River basin within the pristine part of the Bakchar bog was selected as an object of comparison.

2.2 **GIS mapping methods**

Temporal analyses of bog ecosystem dynamics were conducted for the bog contours on the basis of the vegetation maps of key sites. These maps were drawn using the classification of Landsat satellite data using the Semi-Automatic Classification Plugin (QGiS 3.8). The GIS mapping of drained bogs included the following steps: 1) Creating shapefiles, including rectangular contours that limit the territory of key sites. The contour size depended on the properties of the spatial distribution of the plant community within the bogs, the bogs' area and the area of the network of drainage ditches. The area of the Kolpashevo key site is 106 km^2, the Bakchar key site is 44 km^2, the Iksa key site is 25 km^2 and the Ust-Bakchar key site is 4 km^2. 2) The choice of satellite images covers the territory under study. We used an image with a shooting date in the middle of the vegetation period and without clouds. We chose the image from 17 July 2019 for key sites within the Bakchar and Iksa bogs and 31 July 2020 for the Kolpashevo and Ust-Bakchar key sites. 3) A signature list was created with the use of the data of a field study conducted in summer 2019. The key sites were divided into 6–8 classes – forest, field, swamp forest, lags, hummock-hollow bog and pine dwarf-shrub sphagnum drained and pristine bogs. Each of the classes included more than five ROI polygons. 4) The classification of satellite images was conducted with the use of visible and infrared bands (2–7 bands of the Landsat 8 OLI/TIRS) using the maximum likelihood method. The classification results were saved as shapefiles for further analyses.

For analyses of the temporal dynamics of the NDVI, we used the contours of vegetation maps corresponding to pine dwarf shrub sphagnum drained bogs. Landsat satellite data for the period from 1973–2020 were used to assess the state of bogs before the drainage, the effects of the drainage and the modern processes of the self-restoration or degradation of the bogs within key sites. The NDVI was determined using a raster calculator of the Semi-Automatic Classification Plugin. We used data from the Landsat 8 satellite for the period from May–September to identify patterns of the seasonal dynamics of the NDVI for 2019–2020, but to calculate the average values for the growing season by year, we limited ourselves only to summer values (June–August). Due to the high clouds characteristic of the territory under study, the number of images for the period from May–September was only 2 to 8 per year for the period 2013–2020. For an earlier period of shooting (1999–2012, Landsat 7), due to a defect in the equipment, there were even fewer suitable images (1–3 during the growing season); preference was given to images taken from July–August. It should be noted that the continuous series of Landsat 1–5 surveys for the period 1973–1999 is missing, but we managed to obtain data for the period before drainage (1973–1978) and the period in the first years after drainage (1989–1990). In total, satellite data for 77 survey dates were used for analysis, on the basis of which 456 NDVI values for field study points and 252 NDVI values for the vegetation contours of key sites were determined. In order to exclude the influence of NDVI dynamics associated with the peculiarities of the growing season and possible errors associated with the use of a time-heterogeneous data series, a retrospective analysis was carried out using not absolute NDVI values but the ratio of the NDVI values of drained bogs to the NDVI values of the pristine site on the Bakchar bog on each survey date.

 https://doi.org/10.1051/e3sconf/202133302015

3 Results and discussion

3.1 State of vegetation cover and spatial differentiation of the drained raised bogs within key sites

Both pristine and drained pine dwarf-shrub sphagnum plant communities are characterised by limited species of vascular plants due to the oligotrophic conditions of raised bogs, even after drainage. *Pinus sylvestris*, *Chamaedaphne calyculata*, *Ledum palustre* and *Rubus chamaemorus* occurred in all investigated plots. We highlight the change of the species composition of the tree layer within the key site with the greater draining effect only (the Ust-Bakchar bog), evident in the appearance of birch in the pine dwarf-shrub Sphagnum plant community. Also, the height and closeness of the tree layer did not significantly change [1].

The analysis of the NDVI for the vegetation periods from 2019–2020 was carried out to compare the NDVI values with the characteristics of modern vegetation cover. The mean value of the NDVI was 0.70; it varies in the interval 0.68–0.72. The key sites were divided into three groups based on the values of the NDVI of drained bogs relative to the pristine bog (fig. 1). We included the Ust-Bakchar bog in the first group. We explain the higher NDVI values characteristic of both points of the Ust-Bakchar bog by the presence of birch in the species composition of the tree layer and the wide development of the dwarf shrub layer, which reaches a projective coverage of up to 100%. The assumption about the growth of the NDVI being due to the growth of deciduous species was confirmed by analysing the seasonal dynamics of the NDVI. The NDVI values in the Ust-Bakchar bog in May (as of 3 May 2019) are significantly lower (0.47) compared to other sites, where the NDVI varies from 0.55 in the pristine bog to 0.58–0.61 in drained bogs. The second group contains the Kolpashevo bog as well as one part of the Iksa bog. They are characterised by average NDVI values among the key sites relative to the NDVI of a pristine bog. It should be noted that despite the similarity in the NDVI values of the pristine bog and the drained sites at the present stage of vegetation dynamics, the species composition differs between the sites. Probably, in some cases, the NDVI value is achieved due to a more developed tree layer with sparse dwarf shrub and moss cover. The NDVI values are slightly lower (0.68–0.69) in comparison with the pristine bog in the other part of the Iksa bog and the Bakchar bog, which form the third group of sites. In accordance with the results of the cluster analysis conducted on the basis of field data, it was noted that these sites are characterised by the maximum degree of self-restoration of vegetation cover [1]. Thus, an increase in NDVI values relative to a pristine bog is an indicator of the transformation of vegetation cover; with the development of self-restoration processes, the NDVI decreases slightly.

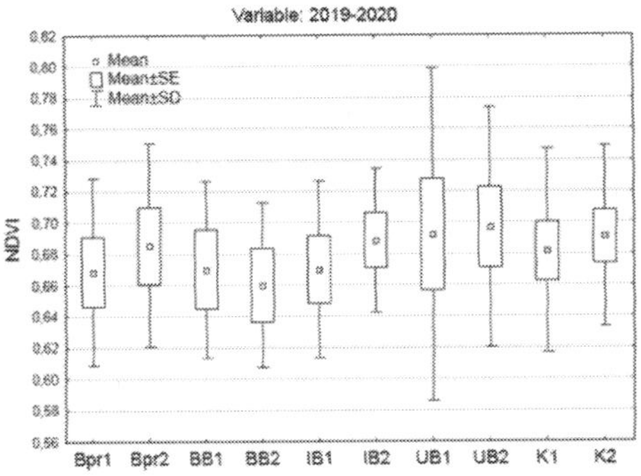

Fig. 1. Diagram with NDVI for all key sites for 2019–2020. See table 1.

3.2 Assessment of temporal dynamics of drained raised bogs

Four stages of the vegetation dynamics of drained bogs were determined on the basis of temporal analyses of the NDVI. Similar patterns of NDVI dynamics are characteristic of most sites (mainly the Bakchar, Iksa and Kolpashevo bogs) (fig. 2). At the first stage, including the period before drainage, the NDVI values were close to those of the pristine site, which confirms that there were similar conditions before drainage. The second stage from 1989–2001(2002) is characterised by a decrease in the NDVI relative to that of the pristine bog. This was probably due to the degradation of moss vegetation, which is a sensitive indicator of a decrease in the water table level in the absence of a significant growth of the tree layer. Furthermore, since the 2000s, there has been an increase in NDVI values and they have stabilised at the level of a pristine bog. The Ust-Bakchar bog is characterised by a more significant decrease of the NDVI in the first years after drainage. This is probably due to a more significant degradation of the moss cover due to a larger decrease in the water table level. By 1999–2000, the NDVI had recovered to the level of a pristine bog. The earlier increase in the NDVI in comparison with the rest of the plots is probably due to the intensive growth of the dwarf shrub layer, in particular, the increase in the cover of *Vaccinium uliginosum*, which is an indicator of drainage. From 2002–2010, we again observed a decrease of the NDVI, which is probably due to a fire that occurred during this period. After 2010, the NDVI values stabilise at a level above those of the pristine bog, which has already been noted above.

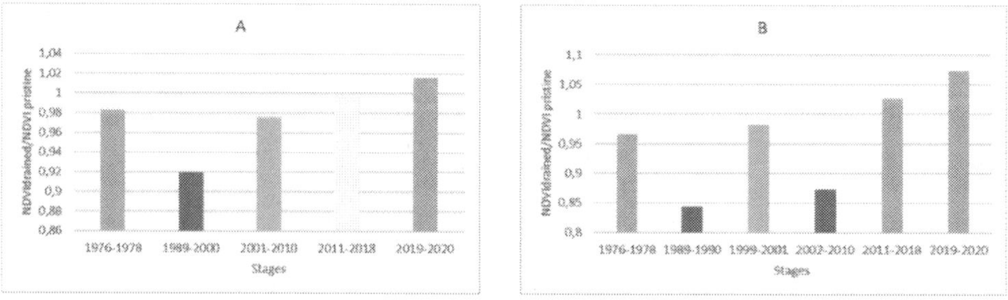

Fig. 2. Temporal dynamics of the vegetation of drained bogs. A – average for key sites (except Ust-Bakchar bog); B – Ust-Bakchar bog.

3.3 Application of Landsat data in wetland dynamics research

Many studies have shown that Landsat data can be used to model the characteristics of wetland vegetation at different stages of recovery; this is an inexpensive and effective method for estimating the carbon storage in the wetlands of different biogeographic regions. At the same time, the authors emphasise the small number of complex studies on the analysis of the spectral characteristics of wetlands and the complexity of mapping wetlands in comparison with forests due to their high spatial heterogeneity, temporal dynamics and high surface humidity. Mapping wetlands requires the development of approaches for the geoinformation mapping of wetlands using satellite data [3, 6].

Knox et al. [12] evaluated the feasibility of using Landsat data as a simple and low-cost tool to assess the balance of recovering wetlands (California, USA). The study showed that plant indices obtained from Landsat data can be used to model photosynthesis within wetlands. Methods for the long-term monitoring of wetlands based on the analysis of multi-temporal images are presented in [4–9]. The authors tested the possibility of using satellite data to assess the characteristics of wetlands' vegetation cover in different regions of the world and developed time series of biophysical characteristics for long-term trend analysis and seasonal dynamics, including the NDVI in their analysis. Amani et al. [3] studied wetland classes using a combination of field and satellite data (Canada). That study concluded that the near infrared and red channels are the most useful spectral ranges for the differentiation of wetlands classes, and they have a high potential for monitoring wetlands.

4 Conclusions

Thus, at most key sites, we did not detect the expected increase in the NDVI which must be happened due to an increase in the production of trees and dwarf shrubs in response to a water table drawdown. On the contrary, as a result of drainage, vegetation degradation has been observed, leading to a decrease in the NDVI in the first years after drainage. Since the beginning of the 2000s, there has been an increase in NDVI values and they have stabilised at the level of a pristine bog.

A more significant transformation of the vegetation cover is characteristic of the Ust-Bakchar bog, which was drained for the purpose of peat extraction. In addition, successional changes of vegetation under the influence of a decrease in the water table level are influenced by fires that periodically occur in this bog as a result of a constantly low water level. For key sites drained for forestry, the effect of drainage was less significant, and currently the NDVI values are close to those of a pristine bog.

Funding: The research was carried out with financial support from the RSF in the framework of the scientific project No. 19-77-00010.

References

1. A. Sinyutkina, Catena **205** (2021). №. 105464.
2. A. A. Sinyutkina, L. P. Gashkova, N. G. Koronatova, A. A. Maloletko, N. P. Mironycheva-Tokareva, I. V. Russkikh, O. V. Serebrennikova, E. B. Strel'nikova, E. K. Vishnyakova Yu. A. Kharanzhevskaya, IOP Conf. Ser.: Earth Environ. Sci. **408**, 012037 (2020)
3. M. Amani, B. Salehi, S. Mahdavi, B. Brisco, ISPRS J. Photogramm Remote Sens. **144**, 119–136 (2018).
4. A. Davranche, G. Lefebvre, B. Poulin, Remote Sensing of Environment **114**, 552–562 (2010)
5. S. Ghosh, D. R. Mishra, A. A. Gitelson, Remote Sensing of Environment **173**, 39–58 (2016)
6. C. Ludwig, A. Walli, C. Schleicher, J. Weichselbaum, M. Riffler, Remote Sensing of Environment **224**, 333–351 (2019)
7. C. Petus, M. Lewis, D. White, Ecological Indicators **34**, 41–52 (2013)
8. L. M. Rebelo, C. M. Finlayson, N. Nagabhatla, J. of Environmental Management **90**, 2144–2153 (2009)
9. C. Vittorio, A. Georgakakos, Remote Sensing of Environment **204**, 1–17 (2018)
10. E.A. Dyukarev, M.N. Alekseeva, E.A. Golovatskaya, Exploring the Earth from space **2**, 38–51 (2017)
11. E.D. Lapshina Mire vegetation of the South-East of Western Siberia. NSU, Novosibirsk (2010)
12. S. H. Knox, I. Dronova, C. Sturtevant, P. Y. Oikawa, J. H. Matthes, J. Verfaillie, D. Baldocchi, Agricultural and Forest Meteorology **237-238**, 233-245 (2017)

E3S Web of Conferences **333**, 02016 (2021) https://doi.org/10.1051/e3sconf/202133302016
RPERS 2021

About disappearance of a large Lake Ebi-Nur, Xinjiang, CHINA

Alexey Terekhov[1,2*], *Nurlan* Abayev[2,3]

[1] Institute of information and computing technology MES, Almaty, 050010, Kazakhstan
[2] RSE Kazhydromet, Almaty, 050022, Kazakhstan
[3] al-Farabi Kazakh National university, Almaty, 050040, Kazakhstan

Abstract. This research considers the process of degradation of the large, bitter-salty Lake Ebi-Nur, located in the Xinjiang (Northwest China), on the border with Kazakhstan. Satellite monitoring using LANDSAT-8 and SENTINEL-2 in recent years has shown the continuation of the process of rapid shallowing of this water body. The area of the lake water mirror decreased from 900 km^2 in 2018 to 210 km^2 in 2021. The reasons for the degradation of Lake Ebi-Nur are the withdrawal of a significant amount of river's runoff for arable land irrigation, against the background of three (2019-2021) strongly low-water years. The stability of the water mirror area of Lake Ebi-Nur and the cultivation of over 600 thousand hectares of irrigated arable land in its basin is possible only with the alternation of low-water and high-water years. Three sharply low-water years in a row have brought the lake to a critical state. The continuation of low water in 2022, most likely, will transfer this lake to the state of a salt marsh. The degradation of Lake Ebi-Nur carries significant environmental risks. This lake is located in a natural area of very strong winds (Dzungarian Gate). The drained parts of lake's bottom become sources of salt storms in direction to agricultural lands and large settlements, including the capital of Xinjiang, city Urumqi. A definite solution to the problem may be the transfer of additional water resources to the Lake Eb-Nur basin from the neighboring water-surplus valleys of the Ili River or the Kara-Ertis River. However, these rivers are transboundary (China-Kazakhstan) and additional returnable seizures of their water resources to the Lake Ebi-Nur will be hard perceived in Kazakhstan.

1 Introduction

The bitter-salty Lake Ebi-Nur, located near the border with Kazakhstan on the territory of the Xinjiang Uygur Autonomous Region of China, is part of the Balkhash-Alakol group of lakes and occupies a last, southeastern position there, see Fig.1. The area of the lake is unstable and in the last 20 years has fluctuated between 500-900 km^2 [1]. This water body is fed by rivers accumulating water from the Eastern Tien Shan ridges (Boro-Horo Ridge,

* Corresponding author: aterekhov1@yandex.ru

Ukeng Ridge). The largest rivers of this lake basin are the Kuitun River and the Bolo-Tala River. The outflows of all these rivers are completely regulated. Part of the runoff is spent on the water supply of settlements, the largest of which is the city of Kuitun, with a population of over 300 thousand peoples. The other part goes to agricultural irrigation. The northern foothills of the Eastern Tien Shan belong to the southern border of the Dzungarian Plain, which is characterized by a hot climate and exceptional aridity. Crop production in this area is possible only with irrigation of arable land. In the Lake Ebi-Nur basin there are over 600 thousand hectares of arable land used for agricultural crops with moderate water consumption (cereals, cotton, etc.). To improve the water supply of the region, two water channels with a length of about 500 km each were carried out here, which transport water from the large transboundary (PRC-Kazakhstan-Russia) Irtysh River (in the territory of the China it is called the Black Irtysh) [2, 3]. The water from these channels does not directly enter into Lake Ebi-Nur basin, but maintains the water supply of the adjacent basin of Lake Manas. This has a positive effect on the water supply of the entire district, since it has a common agricultural irrigation system. In Central Asia, in conditions of arid climate and lack of humidification, water availability of agricultural lands has a high priority, therefore, the low water content of the season primarily affects the state of final water bodies [4].

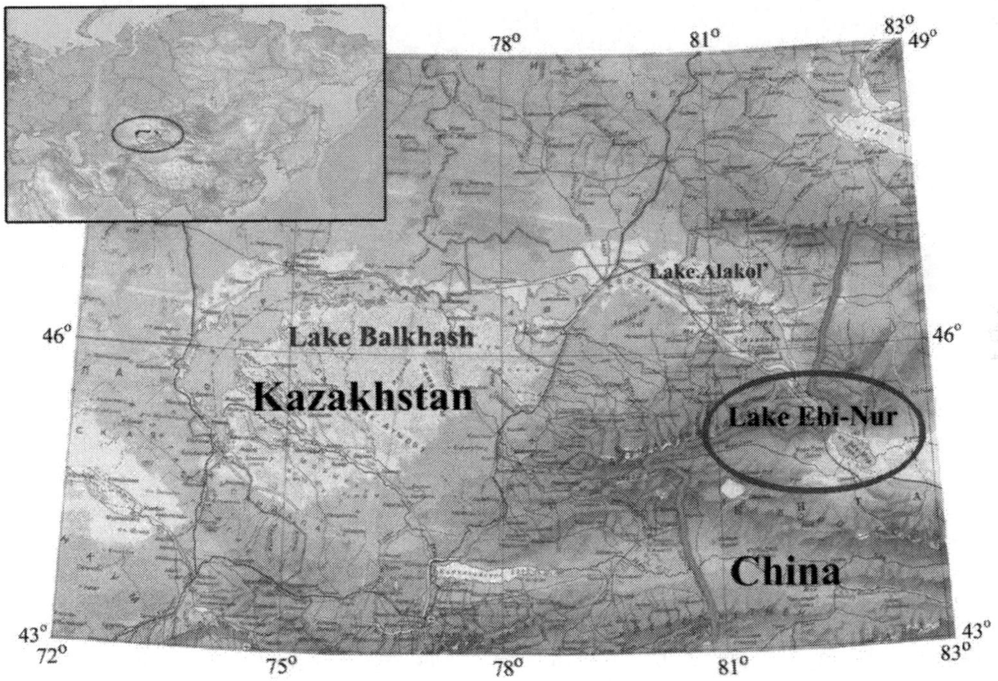

Fig.1. Disposition of Balkhash-Alakol's group of lakes

2 Materials and methods

Lake Ebi-Nur is located in an area characterized by very strong winds (Dzungarian Gate, up to 70 m / sec). The interaction of strong winds with the water mirror of the shallow lake formed a sandbar dividing the lake into two parts. The shallow northern part and the deeper southern part, see Fig.2.

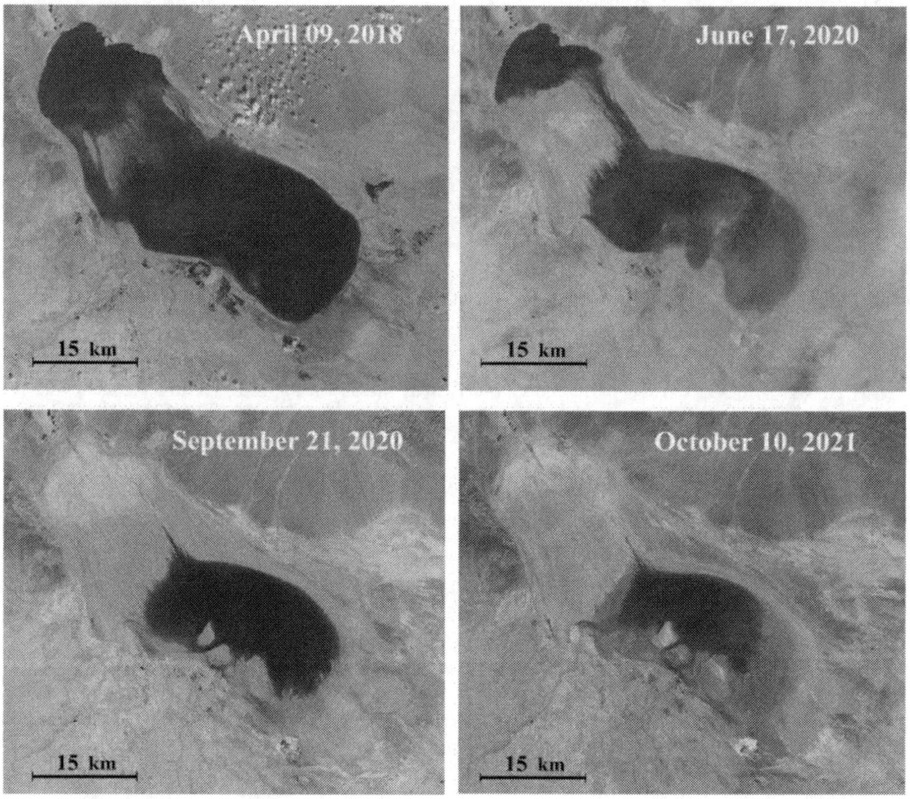

Fig.2. Landsat-8 images of Ebi-Nur Lake during last years. From GloVis (USGS) data portal

Drying up of the northern part of the Lake Ebi-Nur is an extremely unfavorable factor that creates serious risks to the ecology of the foothills of the Eastern Tien Shan. Very strong winds, typical in this area, creates salt storms from the dried-up northern part of the lake bottom. Salt storms carry significant amounts of salt over a distance of up to 100-200 km, which negatively affects the entire ecosystem of the territories adjacent to the lake from the east Lake Ebi-Nur [5], fig.3.

Satellite monitoring of the water mirror area of Lake Ebi-Nur with help LANDSAT and SENTINEL-2 images was carried out earlier [1]. In this study, one more year of observations (2021) was added to the available data. However, this year the condition of the lake has deteriorated dramatically and there is a threat of turning it into a salt marsh.

3 Results and Discussion

According to satellite monitoring (LANDSAT, SENTINEL-2A), the hot and low-water year 2020 led to a significant shallowing of Lake Ebi-Nur [1]. The area of its mirror fell to 340 km^2, a value close to the multi-year minimum of the 1990-2020 period of 339.3 km^2 dated on September 20, 2014. The lack of water in 2021 lowered this value even lower, to 210 km2, which is the absolute historical minimum of the period of satellite observations

(1989-2021), see Fig.4. The continuation of the low-water regime of the rivers of the Eastern Tien Shan in 2022 may transfer the vast majority of the remaining water body of Lake Ebi-Nur to the state of a salt marsh. The formation of a salt marsh is recognized on satellite imagery through the loss of a high spectral difference between the water mirror and land and the appearance of another intermediate spectral class, "not water", but also "not land". In the Landsat-8 image for October 10, 2021 (Fig. 2), a significant part of the lake bottom is already occupied by this new spectral class.

The long-term dynamics of weather regimes in the Central Asian region is characterized by a certain persistence, "long memory". The weather trends are maintained [6,7]. Therefore, there is an increased probability of continuation of the low-water regime in 2018-2021 and in the 2022 season. If the river's outflow in the 2022 season remains low, salt storms, that stopped at the end of the last century due to the flooding of the northern part of Lake Ebi-Nur, may resume in the region. The negative consequences of such a large-scale wind salt transport from the drained part of Lake Ebi-Nur bottom for the ecology of the region as a whole and for agricultural production in the Dzungarian plain are obvious, see Fig.3.

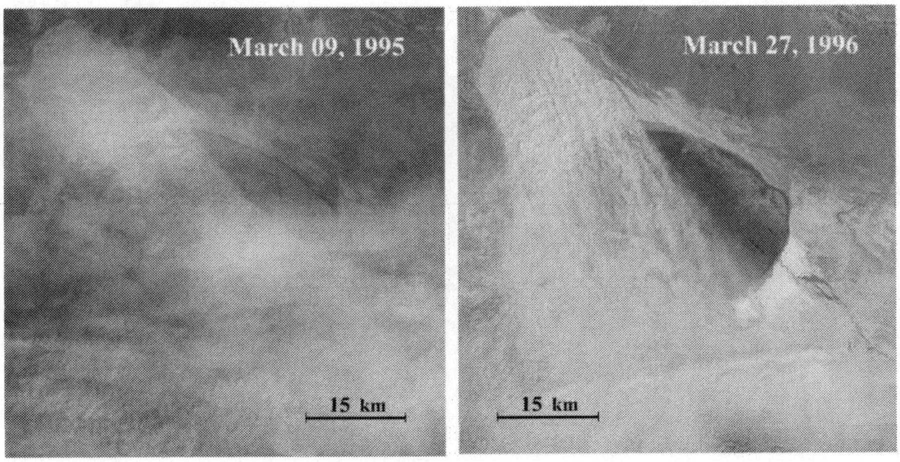

Fig.3 Salt storms. Landsat-5 images of Lake Ebi-Nur during 1995-1996 years. From GloVis (USGS) data portal

4 Conclusion

According to satellite monitoring (LANDSAT-8, SENTINEL-2A), water deficit during 2018-2020 years reduced the area of the seasonal minimum of the Ebi-Nur water mirror (September), to the state of a multi-year minimum. The acute lack of water in 2021 has significantly worsened the situation. The first time in its modern history, Lake Ebi-Nur has approached the state of losing most of the water mirror and turning into a salt marsh. From the south with the Lake Ebi-Nur basin is adjacent to the valley of the big transboundary (China-Kazakhstan) River Ili. This is the largest river in the Xinjiang China, with an average annual outflow at the China-Kazakhstan border is equal 12 km^3. They are separated by a narrow and high ridge Boro-Horo (4359 m), belonging to the Eastern Tien Shan. A technical water pipeline with a length of 50-60 km under the Boro-Horo mountain's range can be created, which will solve the problem of stability of the water mirror area of Lake Ebi-Nur. However, the political aspects of creating such solution are very hard.

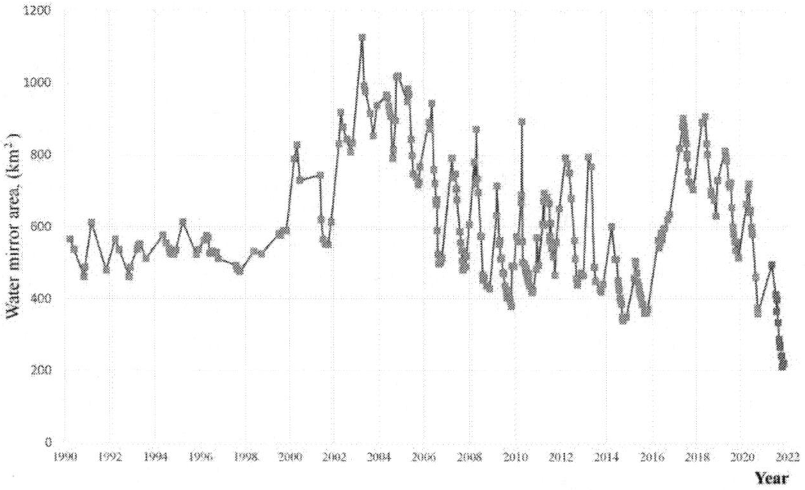

Fig. 4. Satellite monitoring of water mirror area of Ebi-Nur Lake during 1990-2021 years (Landsat 5,7,8; Sentinel-2). Red color is 2021 data. From GloVis (USGS) data portal

Acknowledgment

This research is funded by the Science Committee of the Ministry of Education and Science of the Republic of Kazakhstan (Project No. BR10965172).

References

1. A. Terekhov, N. Abayev, Sovremennye problemy distantsionnogo zondirovaniya Zemli iz kosmosa, **17**, 7, 265-269, (2020), DOI: 10.21046/2070-7401-2020-17-7-265-269.
2. A. Terekhov, Gidrometeorologiya i Ekologiya, 2017, **87**, 4, 54-62, (2017).
3. A. Terekhov, Abst.16-th Conf. "Modern problems of remote sensing of the Earth from space", Moscow, ISR RAS, November 12-16, 117, (2018), DOI: 10.21046/2070-16DZZconf-2018a.
4. A. Terekhov, N. Abayev, E3S Web of Conference, **223**, 02009, (2020), DOI: 10.1051/e3sconf/202022302009.
5. J. Abuduwaili, M. Gabchenko, X. Junrong, Arid. Environ **72**, 10, 1843-1852, (2008), DOI: 10.1016/j.jaridenv.2008.05.006.
6. A. Terekhov, I. Vitkovskaya, N. Abayev, E3S Web of Conference, **149**, 03004, (2020), DOI: 10.1051/e3sconf/202014903004.
7. R. Dole, Monthly Weather Review, **114**, 1, 178–207, (1986), DOI: 10.1175/1520-0493(1986)114<0178:PAOTEN>2.0.CO;2.

Satellite data in thermal range for natural and technogenic ecosystems monitoring

Nikita Yakimov[1,2*], *Evgenii* Ponomarev[1,3], and *Tatiana* Ponomareva[1,3]

[1]Siberian Federal University, Krasnoyarsk, 660041, Russian Federation
[2]Federal Research Center «Krasnoyarsk Science Center SB RAS», Krasnoyarsk, Russian Federation
[3]V.N.Sukachev Institute of Forest, Siberian Branch, Russian Academy of Sciences, Krasnoyarsk, 660036, Russian Federation

Abstract. A method for recovery monitoring in post-fire and post-technogenic landscapes was proposed based on satellite data in a wide spectral range, including the infrared band data. A decrease in the spectral surface albedo in post-fire areas, caused by the destruction of on-ground vegetation, provokes excessive heating of the surface and upper soil layer. Surface thermal anomalies were evaluated under conditions of changes in the heat-insulating properties of vegetation and ground cover. The relative temperature anomalies in post-fire plots (overestimation up to 30% compared to non-disturbed territory) are typical for permafrost conditions of Siberia. Similar process was recorded for both natural (post-fire) and post-technogenic landscapes. Within 22 years after the fire, thermal insulation properties of the vegetation cover were restored. Thus, the relative temperature anomaly (of 3±1%) has reached the background value. In post-technogenic plots, conditions are more "contrast" compared to the background, and restoration of the thermal regime takes significantly longer (>60 years). "Neo-technogenic ecosystems" with specific soil thermal regimes compared to the background ones are formed both for reclaimed and for non-reclaimed post-technogenic plots. On average, surface temperature has overestimated at least by 10–15% in post-technogenic plots compared to non-disturbed territory.

1 Introduction

Nowadays, long-term forest destruction monitoring in Siberia is impossible without satellite monitoring tools [1, 2]. Thematic processing of multispectral satellite data makes it possible to identify vegetation damage and evaluate seasonal or intraseasonal dynamics of disturbed vegetation and soils. Thus, modern approaches based on remote sensing data make it possible to control and predict the state and successions in natural or technogenic ecosystems. Satellite monitoring allows estimating the annual increase of vegetation disturbances in Siberia, which are associated with a number of destructive natural (insect pests, wildfires, other natural processes) or human-made factors (logging, activities of the mining complex, etc.).

Disturbance of heat-insulating properties of the soil and the vegetation cover is the reason for the change in thermal regime within local areas. When such changes accumulate, they should be considered a significant factor regulating the state and dynamics of ecosystems and soil and permafrost characteristics [3–6]. Up to 20% of permafrost larch forests in Siberia [7, 8] are subject to post-fire changes in vegetation and surface and soil temperature regimes. Anomalous heating of the surface and the soil layer is also observed in post-technogenic plots (open-pit mining, quarries, overburden dumps, loggings, etc.) under the strong mechanical impact on vegetation cover and soil. Landscape thermal regime monitoring and the development of remote sensing techniques in monitoring recovery processes are urgent. The issue can be effectively solved using satellite data in a wide spectrum range, including data in the infrared (IR) range.

The destruction of on-ground cover affects surface albedo, emissivity, moisture, and water regime of the upper soil horizons. As a result, the temperature regime of soils changes significantly compared to the background areas. Such thermal anomalies of surface and soil can remain significant for a long time in post-fire plots [5–7] and post-technogenic areas when the ground cover and soils are disturbed mechanically [9, 10].

The main aim of the study was to compare long-term peculiarities of thermal anomalies in post-fire and post-technogenic plots in Siberia based on Landsat survey data in the infrared (IR) range. In addition, using time series of Terra/MODIS satellite data, we performed indirect assessments of disturbed ecosystems recovery in terms of disappearing the spectral features anomaly (spectral albedo in the short-wavelength bands and vegetation index) compared to characteristics of non-disturbed background areas.

* Corresponding author: nyakimov96@mail.ru

2 Initial data

For technogenic ecosystems, we used satellite images of average spatial resolution (15–60 m) from Landsat 2–8 for 1975–2021, which are freely available in The United States Geological Survey (USGS) database (https://earthexplorer.usgs.gov/, accessed on 30 October 2021). The surface temperature was evaluated from the calibrated B6 channel (λ = 10.4–12.5 μm, Landsat-5/TM – Thematic Mapper), B6/1 channel (Landsat-7/ETM – Enhanced Thematic Mapper), and B10 channel (λ = 10.6–11.9 μm, Landsat-8/OLI – Operational Land Imager). Using calibration constants from the metadata files, we implemented the radiometric correction method to the initial data [11]. Calibrated data from the NIR (λ = 0.63–0.69 μm) and RED (λ = 1.55–1.75 μm) channels were used to calculate Normalized Difference Vegetation Index (NDVI).

For natural ecosystems, we used low spatial resolution survey data (250–1000 m) from Terra/MODIS (Moderate Resolution Imaging Spectroradiometer) for 1996–2018. Standard products L2G and L3 used (free USGS database, https://lpdaac.usgs.gov/dataset_discovery/modis, accessed on 30 October 2021). To analyze NDVI and spectral albedo, we operated with reflectances measured in MODIS bands #1 ($\lambda 1$ = 0.620–0.670 μm) and #2 ($\lambda 2$ = 0.841–0.876 μm) (standard product MOD09GQ). To analyze surface temperature anomalies, we used the MOD11A1 standard product. We used 10-day averaged data across the entire initial data set, focusing on the disturbed ecosystem's recovery succession stages. A monthly data average procedure and a procedure of spatial averaging within each post-fire or post-technogenic plot were applied.

The relative NDVI and temperature field anomalies were defined in percentage compared to background ones (percent relative to the average long-term norm).

3 Methods and data processing

In the current research, we analyzed data for typical taiga region of Central Siberia (59–66° N, 90–107° E). Zonal and intrazonal soil types represent the soil cover of the study area. The forest-steppe part of the territory is characterized by full-profile soils with a well-differentiated profile, as well as with high humus content, and a heavy granulometric composition. Moderately thick chernozems primarily represent soils. The soils in the middle taiga zone dominated by podzols. Peat, gley, and humus soils are typical in areas with poor drainage. Parent rocks determine the genetic specificity of the soils in the permafrost zone. The soil depth is low (0.20–0.50 m) here, and the soil profile's vertical differentiation is weak. The permafrost level (<1.0 m) depends significantly on the topography [12]. Permafrost is widespread on leveled elements and not be present in the soil profile on drained areas of slopes and valleys. Periodic freezing and thawing of the soils determine the soil structure, the low rate of humus accumulation in the soil, the peculiarities of moisture transport, waterlogging. Moreover, freezing and thawing processes affect slightly acidic or neutral soil reaction and loamy or clayey granulometric composition with a high gravel content in the lower part of the soil profile [13].

Wildfires affect soil properties [12, 17] and change the soil structure [4, 14]. After a wildfire, the soil loses upper horizon, the level of permafrost occurrence decreases, and the soil formula transforms to: Opir-O(F + H)pir-CR-C. Under extreme heat stress, the profile structure can transform more significantly. Depending on the degree of litter burnout, the soil formula changes to: O(F + H)pir-CRpir-C or O(H)pir-CRpir-C. Locally, complete burnout of the upper organic horizons of the soil is possible [14]. As a rule, the upper part of the soil profile has a thin humus-accumulative (underdeveloped) horizon W or a fragmented coarse humus horizon Oao both in case of post-fire mineralization of cryozems and in case of technogenic transforming of soils. Such soils formula is: Oao(W)-C.

Under the technogenic impact (open-pit mining, quarries, overburden dumps) the soil cover significantly transforms. Under the prevailing conditions of soil formation, specific technogenic surface formations (abralites, lithostrats) and anthropogenically transformed types of natural soils are formed. On the territory of the residential zone, urban soils are formed such as urbiquazizems and artindustrates (Technosols).

The general regularities of soil transformation after destructive effects on heat-insulating (thermo-isolation properties) covers (an increase in thawing depth, an increase in soil moisture due to thawing of schlieren and wedge ice, an increase in gleying, a cryoturbation processes) depend on the vegetation and soil restoration succession. The relative temperature anomalies in post-fire plots (PF) are typical for permafrost conditions of Siberia. Similar process was recorded for both natural (post-fire) and post-technogenic landscapes [4]. Thus, the effect of a post-fire increase in the thawing depth (by 30–100% in the first year after the fire) [4] leads to a unique path of soil evolution, when the underlying gleyed rock layers are included in soil formation for a significant period [15]. This type of evolution is possible only in the permafrost zone. Figures 1–2 show the technogenic objects studied.

Fig. 1. Borodinsky Coal Mine quarries and overburden dumps (Krasnoyarsk Region, Siberia), 2010–2017.

Fig. 2. Olympiada Mining Complex quarries and overburden dumps (Krasnoyarsk Region, Siberia), 2019.

We selected satellite data for the summer growing season (July), making it possible to trace sporadic maxima of heat anomalies. For post-fire plots investigating we analyzed data on 10–15 fires registered annually in Siberia. Next the spectral anomalies dynamics were analyzed for two technogenic sites: the Borodinsky Coal mine (BCM) for 1955–2018 and the Olympiada Mining Complex (OMP) for 1989–2019.

The satellite data time series allowed us to estimate the disturbed land restoration in terms of spectral and thermal anrelative anomalies compared to non-disturbed background areas. We assessed long-term changes in anomalies of spectral characteristics (spectral albedo and surface temperature) and the state of vegetation cover (NDVI) compared to background areas. The period of stabilization of the properties of disturbed ecosystems was evaluated as well. Using ArcMAP GIS software we implemented the procedure of temporal and spatial average of a large amount of data. We estimated the anomalies of albedo, surface temperature, and NDVI during 1, 5, 10–12, 20–22, 40, 55–60 years after destructions in percentage compared to the long-term statistical norm. Figures 3–5 demonstrate different stages of recovering in terms of thermal fields.

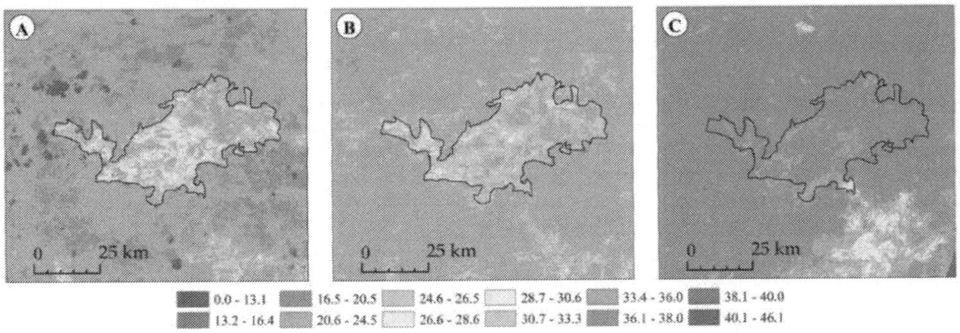

Fig. 3. Dynamics of surface temperature in post-fire plot (PF) burned in 1997 after the 1st year of recovery (A), 10-years recovery in 2006 (B), and 20-years recovery in 2016 (C). Data from Landsat imagery in thermal range.

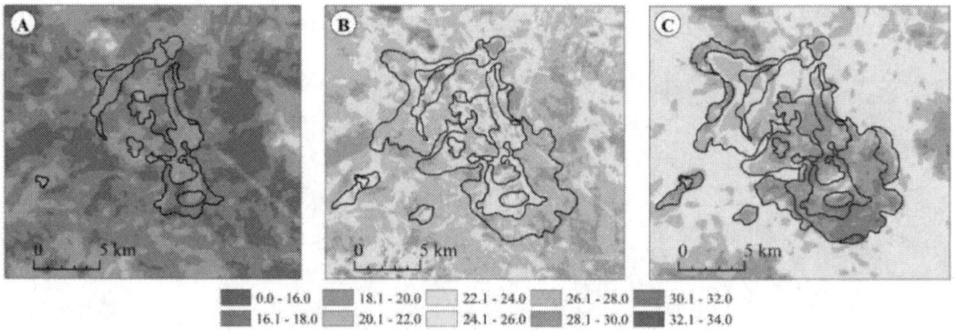

Fig. 4. Dynamics of surface temperature in OMP plot with the identification of different-aged zones of a technogenic ecosystem formation (since 1989) on dumps in 1989 (A), 2000 (B), and 2019 (C). Data from Landsat imagery in thermal range.

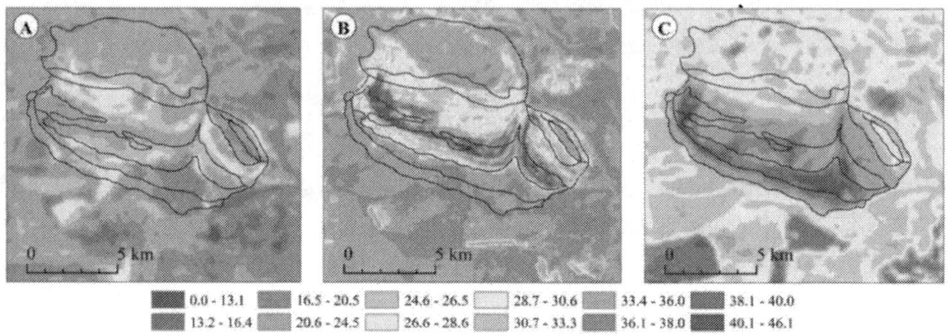

Fig. 5. Dynamics of surface temperature in BCM plot with the identification of different-aged zones of a technogenic ecosystem formation (before 1975) on dumps with reclamation in 1990 (A), 2000 (B), 2018 (C). Data from Landsat imagery in thermal range.

The relative anomalies of the surface temperature we evaluated as ratio of absolute values of disturbed plots overestimation to the temperature of background territory ($\Delta T/T_{bg}$) in percent. The same procedure was applied for estimating of spectral albedo anomalies ($\Delta\alpha/\alpha_{bg}$) and NDVI anomalies ($\Delta NDVI/NDVI_{bg}$). Based on the data obtained, we constructed the dynamics of temperature anomaly and NDVI index after the impact of different factors.

4 Results and Discussion

A decrease in the spectral albedo in post-fire (fire damage) and post-technogenic areas (due to mechanical damage to the surface soil layer [16]) provokes excessive heating of the underlying surface, which persists for decades. This effect occurs in the summer, late spring, and early autumn at positive air temperatures. It was found that after fires, surface temperature changes by an average of up to 33% compared to the background. The maximum recorded rise was 46%. Surface temperature is rising due to the high degree of suppression of vegetation [17]. For at least 10 years, the level of anomalies decreases to 11%, with a maximum of 20%, while for at least 20 years, the amplitudes of anomalies decrease to the level of background values and measurement errors in the range from 3–5%. When analyzing the data on temperature of post-technogenic areas, it was revealed that high values of thermal anomalies from 70 to 100% in the initial period gradually decreases to 19–20% after at least 40 years of recovery. A greater rate of decline in the temperature anomaly is observed during the first 10 years (maximum excess decreases by 35%). Then, over 40 years, there is a decrease in the maximum fixed anomalies by about 15% per 10 years. The value of the thermal field anomaly does not decrease so intensively (up to ~10–13% per 10 years) 40–60 years after the disturbances. So temperature parameters are assumed to be stabilized in this time interval. Anomalies and dynamics of spectral albedo recovery depend on the reclamation procedures in the post-technogenic area. Notably, such a high level of a relative anomaly in technogenic areas persists twice as long as in areas prone to fires, where recovery occurs naturally [17]. Long-term temporal-series analysis showed that in post-technogenic ecosystems, the properties of the upper soil horizons did not recover to the background level for a long time, and thermal regimes remained significantly anomalous for more than 60 years after technogenic impact. Thus, we suggest the formation of "neo-technogenic ecosystems" with specific thermal regimes of soils that differ from the background ones.

The dynamics of relative vegetation index anomalies of disturbed plots compared to the background ($\Delta NDVI/NDVI_{bg}$) were evaluated for the PF plots and post-technogenic plots with reclamation (BCM). Due to a lack of vegetation restoration, dynamics were not fixed for non-reclaimed plots (OMP) [12, 18, 19].

The long-term average NDVI of the background was ~0.6. After the fire (PF plots), the value of $\Delta NDVI/NDVI_{bg}$ reached 50–65% (with sporadic maxima in some plots ~80% depending on the intensity of wildfire). During 10–12 years, the value of the vegetation index anomaly decreased to 15–18% (with sporadic maxima ~26%). After 15–20 years (the time of stabilization (T_{stab}) of the spectral characteristics), the amplitude of anomaly decreased to background values and could be underestimated <5% of background values. We used exponential function to approximate (R^2 = 0.95) the dynamics of $\Delta NDVI/NDVI_{bg}$ (Figure 6, c). It should be noted [5] that the vegetation recovery rate was two times higher than the temperature anomalies recovery rate, which was also approximated by an exponential function (R^2 = 0.85) (Figure 6, a, c).

In the post-technogenic BCM plot, we identified different-aged zones of a technogenic ecosystem formation since 1955. We considered the time series to analyze the dynamics of $\Delta NDVI/NDVI_{bg}$ since the 100% anomaly after technogenic impact (Figure 6, d). Actual data show that until 2000, there was no successful restoration of vegetation. At the same time, the data obtained after 20 years show that the zones developed in a later period and reclaimed with the planting of coniferous seedlings have a more intensive vegetation restoration [16, 19]. Thus, $\Delta NDVI/NDVI_{bg}$ anomaly decreased up to 33–48% during 30 years, similar to the values for PF plots after only ~5 years of natural recovery.

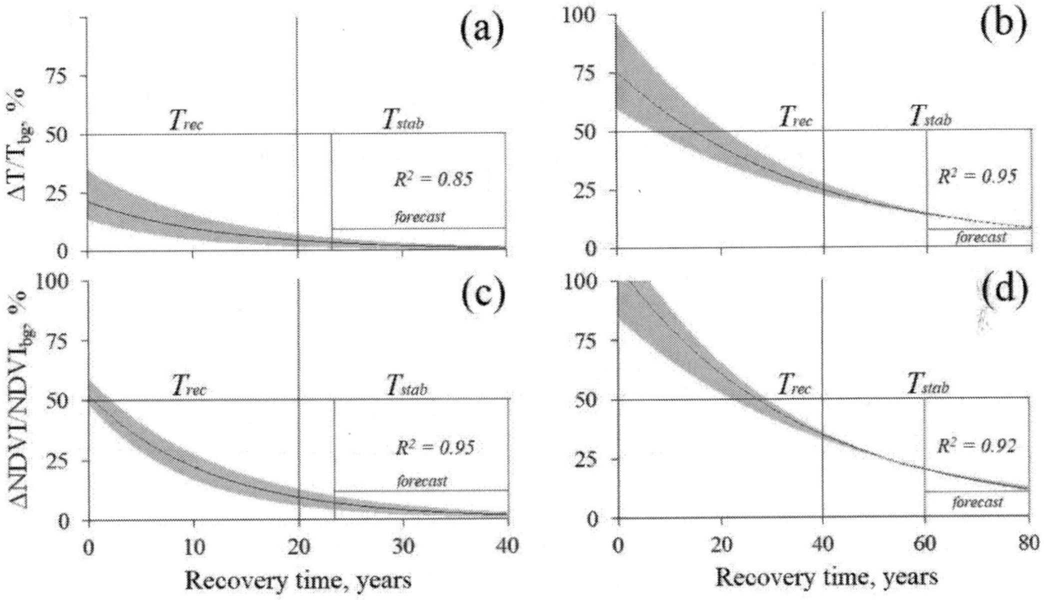

Fig. 6. Long-term dynamics of the relative anomaly in post-fire areas (a, c) and BCM plots (b, d). T_{rec} is a period of intensive recovery, T_{stab} is a period of stabilization of parameters. Standard Deviation is coloured in gray.

The peculiarity of plots after fire disturbances is that natural recovery processes occur there. The peculiarity of post-technogenic plots is that there are two possible options for ecosystem dynamics: 1) restoration after reclamation or 2) long-term condition in the format of non-reclaimed lands (dumps, mineralized surfaces). Each of these options plays a significant role in forming surface temperature anomalies and their effect on the properties of soil horizons [4, 12, 17–19].

5 Conclusions

Under conditions of the same insolation, plots with disturbances in the upper soil horizons and ground cover are accompanied by the formation of long-term surface temperature anomalies. Similar processes were recorded for both natural (post-fire) and post-technogenic landscapes. Within 20 years, the thermal insulation properties of the vegetation cover restore in the post-fire areas. Thus, the relative temperature anomaly reaches the level of the mathematical error of measurement. In post-technogenic plots, conditions are more "contrast" compared to the background, and the restoration of the thermal regime takes longer time (>60 years). "Neo-technogenic ecosystems" are formed, characterized with special thermal regimes of soils that differ from the background ones both for reclaimed (BCM) and for non-reclaimed (OMP) plots. Thus, surface thermal anomalies monitoring can be used as an additional diagnostic criterion of post-technogenic ecosystems state in the context of territories (and ecosystems) recovery after technogenic impacts.

Using time series of satellite data (Landsat and Terra/MODIS), indirect assessments of disturbed ecosystems recovery is possible in terms of disappearing the spectral features anomaly (spectral albedo in the short-wavelength bands and NDVI) compared to characteristics of the background areas.

Remote sensing data make it possible to effectively monitor the state of disturbed territories on the scale of regions and natural zones. The proposed approach, based on the use of the thermal range, extended the limit of remote monitoring of the state of disturbed ecosystems to 20–40–60 years with the ability to control and predict the dynamics of recovery processes based on periodic satellite infrared surveys of the territory.

Acknowledgments

The present study was performed using the subject of project of IF SB RAS no. 0287-2021-0010. This research was partly funded by the Russian Foundation for Basic Research (RFBR) and Government of the Krasnoyarsk krai, and Krasnoyarsk krai Foundation for Research and Development Support, no. 20-44-242002 ("Instrumental monitoring of physical properties and numerical modeling of the state of technogenically disturbed soils in Siberia").

References

1. S.A. Bartalev, F.V. Stytsenko, V.A. Egorov, E.A. Lupyan, Lesovedenie, **2**, 83-94 (2015)
2. V.G. Bondur, K.A. Gordo, V.L. Kladov, Izvestiya, Atmos. Ocean. Phys., **53**, 859-874 (2017)
3. A.A. Knorre, A.V. Kirdyanov, A.S. Prokushkin, P.J. Krusic, U. Buntgen, Sci. Total Environ, **652**, 314-319 (2019)
4. L. Orgogozo, A.S. Prokushkin, O.S. Pokrovsky, C. Grenier, M. Quintard, J. Viers, S. Audry, Permafrost and Periglac Process, **30**, 75-89 (2019)
5. A. Kirdyanov, M. Saurer, R. Siegwolf, A. Knorre, A.S. Prokushkin, O. Churakova, M.V. Fonti, U. Büntgen, Env. Res. Let., **15**, 034061 (2020)
6. E.I. Ponomarev, O.V. Masyagina, K.Y. Litvintsev, T.V. Ponomareva, E.G. Shvetsov, K.A. Finnikov, Forests, **11**, 790 (2020).
7. E.I. Ponomarev, T.V. Ponomareva, Contemporary Problems of Ecology, **11**, 420-427 (2018)
8. V.I. Kharuk, E.I. Ponomarev, G.A. Ivanova, M.L. Dvinskaya, S.C.P. Coogan, M.D. Flannigan, Ambio, **50**, 1953-1974 (2021)
9. K.V. Krasnoshchekov, A.V. Dergunov, E.I. Ponomarev, Current problems in remote sensing of the Earth from space, **16**, 87-97 (2019)
10. Ł. Uzarowicz, P. Charzyński, A. Greinert, P. Hulisz, C. Kabała, G. Kusza, W. Kwasowski, A. Pędziwiatr, Soil Science Annual, **4**, 281-299 (2020)
11. N. Mishra, Md.O. Haque, L. Leigh, D. Aaron, D. Helder, B.L. Markham, Remote Sens, **6**, 12619-12638 (2014)
12. A. Dymov, E. Abakumov, I. Bezkorovaynaya, A. Prokushkin, Y. Kuzyakov, E. Milanovsky, Theoretical Ecology, **4**, 13-23 (2018)
13. A.P. Abaimov, A.I. Bondarev, O.A. Zyryanova, S.A. Shitov, Nauka, 705 (1997)
14. A.A. Dymov, Y.A. Dubrovsky, D.N. Gabov, Eurasian Soil Sc, **47**, 47-56 (2014)
15. N.S. Mergelov, Izv. Ross. Akad. Nauk, Ser. Geogr, **3**, 129-140 (2015)
16. T.V. Ponomareva, N.M. Kovaleva, A.S. Shishikin, E.I. Ponomarev, Gornyi Zhurnal, **10**, 48-53 (2020).
17. I.N. Bezkorovaynaya, I.V. Borisova, A.V. Klimchenko, O.M. Shabalina, L.P. Zakharchenko, A.A. Ilyin, A.K. Beskrovny, Bulletin of KSAU, **9**, 181-189 (2017)
18. S.G. Kornienko, Earth's Cryosphere, **21**, 93-104 (2017)

Identification of genetic soil types based on remote sensing data

Ekaterina Kulik, *Arslanbek* Bajshuakov

Siberian State University of Geosystems and Technologies, Photogrammetry and Remote Sensing Department, 10 Plakhotnogo street, 630108, Novosibirsk, Russia

Abstract. At the present stage of social development traditional contact methods of studying the territory features are inferior to other, less costly and more profitable technologies. Such technology is automation of space imagery information interpretation. It allows to minimize scope of field works and it is economically and managerially profitable method of territory analysis. So, in the work the methods of digital analysis and automatic Earth remote data sensing interpretation are used for identification of genetic soil types. During the research conditions of the land has been determined by calculation and index images analysis. The application of interpretational signs had allowed to identify genetic soil types. The methodology and the results of the analysis can serve as a basis for solving the problems of updating soil maps.

In the tasks of monitoring natural resources, it is very important to identify the characteristics of soils. Soil fertility assessment is considered an integral part of agricultural monitoring. The analysis of the soil condition makes it possible to identify degraded and disrupted lands in order to further develop measures for soil recovery. By analyzing the remote sensing data (RSD) of the Earth, it becomes possible to monitor soil fertility, create and update soil maps. Their use helps agricultural producers to carry out soil suitability classification and soil rating. The purpose of this study is the application of methods for automating the RSD analysis for finding-out of genetic soil types.

To accomplish purpose, there were formulated tasks aimed at in-depth geographical exploration of territory. The tasks are to determine of vegetation and soil conditions using the SAVI soil index and to map soil types identified by interpretational signs.

As source data work uses multispectral image received by SPOT-6 satellite system. The SPOT-6 spacecraft engineered by Airbus Deference and Space (France). The acquisition is performed from the orbit that has an altitude of 694 km. Each SPOT-6 scene is about 60 kilometers long and 60 kilometers wide. Spectral bands of imaging instruments are presented by optical wave channels in the range from 0.45 to 0.69 μm as well as near-infrared channel in the range from 0.76 to 0.89 μm. For panchromatic acquisition the range from 0.45 to 0.75 μm was used [1]. The soil map from 1985 was used as additional thematic information. The study area is located in the northern part of Dovolensky district of Novosibirsk region on the lands of the Utyanka municipality. These lands have been in agricultural use for a considerable number of years.

The topography of the study area is slightly dissected wavy plain with low parallel-oriented from southwest to northeast hill chains up to 15 meters. The hill chains are occupied by chernozem soils [2]. The valleys between hill chains are occupied by wetlands and lakes where boggy and peat-boggy soils are spread. Most of the study area is occupied by forest-steppe plains, which are dominated by solonetzes. Parent materials are loam and sandy loam.

The content in plants is rather heterogeneous. There are grassland steppes and steeped grasslands on solonetzes. In depressions occupied by aspen-birch outlier forest there are

https://doi.org/10.1051/e3sconf/202133302018

solods. On lakeland terraces there are solonchaks with sparse vegetation represented by specific halophytic species.

In relatively small areas fitted within the individual satellite high-resolution imageries it is possible to conduct research to identify seasonal and/or multi-year local changes in the characteristics of geosystems. For soil types interpretation it is advisable to use images of the spring and autumn period, when the vegetation is more sparse or absent at all. The most reliable genetic soil types identification is achieved where feasible to analyze such direct interpretational signs as spectral brightness, color, background, size, shape, structure, texture, and pattern. Indirect interpretational signs of soil types are vegetation and relief features. Also it may be considered the use of such auxiliary material as soil, topographic and geobotanical maps if the high recognizing reliability of objects types from their images is aimed [3].

For identification of areas with bare ground and vegetation the SAVI (Soil Adjusted Vegetation Index) index image calculation method was used [4]. The SAVI index is calculated according to formula 1 using L correction factor. It is introduced to bring the index with sparse vegetation to the NDVI range:

$$\mathrm{SAVI} = \frac{(\mathrm{NIR}-\mathrm{Red})}{(\mathrm{NIR}+\mathrm{Red}+\mathrm{L})} * (1 + \mathrm{L}) \quad (1)$$

The SAVI soil index was calculated using the ScanEx Image Processor software package, in which a value greater than 0.5 is set for the L correction factor if the vegetation cover is not significantly presented on the ground [5]. The most convenient way to work with these indices is to use images before the tillering period of agricultural vegetation, during the period of early germination and after harvesting. After setting the parameters in the SAVI soil index operation in the ScanEx Image Processor software package, a grayscale image is produced in the output. The SAVI color palette is used to color the result, similarly to the NDVI. According to the given color scheme, the resulting index image shows the condition of the vegetation.

The calculation of the vegetation soil index SAVI resulted in the identification of areas with bare soils which are characteristic of arable land located on chernozem soils. As the image of the autumn period was used, a visual assessment confirmed that crops had been harvested. Image analysis had shown that most of the area is occupied by sparse vegetation.

There is a correlation directly between the results obtained on the basis of the analysis of the index image presented in Figure 1 and the soil types of the local territory, since the boundaries of the areas with identified soil types are similar to the contour part of the retrospective soil maps. It confirms the eligibility of use vegetation indicators (the state and characteristics of the vegetation are determined by the values of vegetation indices) in this type of interpretation of remote sensing data.

The lake shoreline corresponds to boggy and peat-boggy soil types. The boundaries of areas with weak and sparse vegetation are similar to the boundaries of areas with meadow and solonetz soils. Areas with dense vegetation are characterized by outlier forests under which solod soils are formed. So identification of genetic soil types and delineation of their extent can be performed using automated analysis of vegetation index images, where vegetation indicators act as reliable indirect interpretational signs.

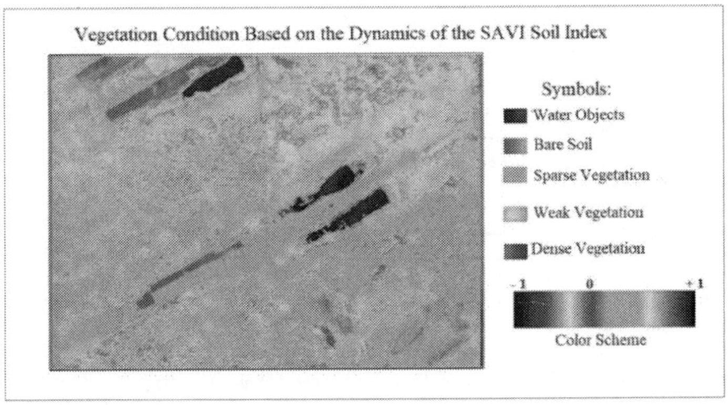

Fig. 1. Index Image of SAVI

Using indirect interpretational signs, vegetation and topography indicators as well as index image and 1985 soil map data, a soil type scheme including eight genetic types was produced in the ArcGIS software package. The following soil types have been identified:
- chernozems,
- meadows,
- boggies,
- meadow-chernozemics,
- chernozem-meadows,
- solonetzes,
- solonchaks,
- solods (Figure 2, Table 1).

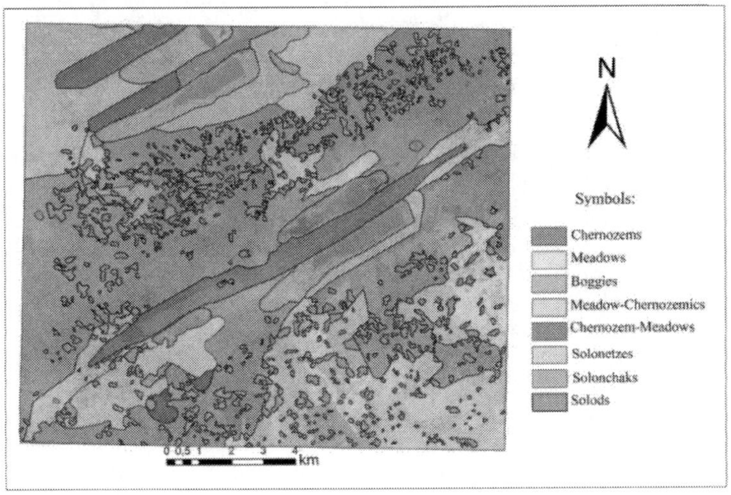

Fig. 2. Scheme of Soil Types

Table 1. Soil types

Name in English	Name in the WRB system
Chernozems	Chernozems
Boggies	Fibric Histosols
Meadows	Eutric Vertisols
Meadow-chernozemics	Luvic Chernozems
Chernozem-meadows	Haplic Arenosols
Solonetzes	Solonetz
Solonchaks	Solonchaks
Solods	Haplic Gleysols/ Solodic Planosols

The chernozem soils are formed in areas with a flat surface, occupied by crops with a high accumulation of organic matter (humus). Black tone, parallel-striped image texture and color are used for interpretation. The chernozem-meadow soils can be formed under croplands. Such soils are created under a highly productive herbaceous cover, and have a higher humus proportion than chernozem soils.

The meadow soils are generally formed under herbaceous vegetation, which serves as an indicator for the identification of this soil type. The interpretational signs are homogeneous structure and color.

The solonetzes are formed on saline rocks on the plains and in the lowlands. Vegetation on solonetz soils is represented by specific species with deep root systems such as artemisia (*Artemísia*), fescue (*Festuca valesiaca*) and *Stipa*. The heterogeneous grain structure, spotty texture and different shades of color have become interpretational signs for the identification of solonetz soils. However, the identification of solonetz soils was accompanied by the availability of data from a retrospective soil map.

As distinct from the solonetzes, solonchaks contain water-soluble salts on the surface. There are easily recognizable visually on the images, having a light tone and homogeneous structure at interpretation. The vegetation on solonchaks is sparse and represented by specific halophyte species.

Solod soils do not occupy continuous vast areas, but they are usually found as local areas within the steppe and forest-steppe zones. These soils are formed in small depressions under the aspen and birch outlier forests.

The boggy and peat-boggy soils are found in depressions on the site of lake and waterlogged basins.

As a result of genetic soil types identification, a soil type scheme was made using the high-resolution satellite image and the soil map of 1985. In creating this scheme, the relationships between the soil types and the indicators for their identification were studied.

Calculation of various index images is a very flexible, quite popular and often universal method of multispectral satellite data analysis to obtain information about the characteristics of vegetation and soil cover [6]. As a result of research on this subject the authors studied geographical features of the area of interest. There was applied methods of digital image processing such as calculating the SAVI index and using interpretational signs for the of soil types mapping. The represented results with identified soil genetic types based on remote sensing data will allow agricultural enterprises to outline rational methods of soil treatment.

This study was conducted with the information support of private Limited Liability Company Engineering and Technology Center "SCANEX" (with the supply of satellite images from the archive). It is based on the validity of the right to receive information support and images by the winners of the All-Russian Student Olympiad on Thematic Earth Remote Sensing Data Interpretation held at the Reshetnev Siberian State University of Science and Technology in August 2017.

Digital satellite image processing was carried out using ScanEx Image Processor software package. The results were processed in ArcGIS software package.

References

1. E. Kulik, A. Bayshuakov, *Monitoring of the agricultural crops states at different vegetation stages using remote sensing data,* Regional Problems of Earth Remote Sensing: Proceedings of the VII International Scientific Conference, 29 September – 2 October 2020, pp. 248 – 250, Krasnoyarsk, Russia (2020)

2. A. Bayshuakov, *Analysis of geographical features from satellite imagery for soil mapping purposes*, Aktual'nye Napravleniya Geograficheskih Issledovanij v Kemerovskoj Oblasti i Drugih Regionah Rossii [Topical Areas of Geographical Research in the Kemerovo Region and Other Regions of Russia], 25–28 February 2021, pp. 84 – 88, Kemerovo, Russia (2021)

3. A. Gordienko, T*he use of index imagery in detecting changes in multi-temporal satellite images*, Interekspo GEO-Sibir' [Interexpo GEO-Siberia], 13 – 25 April 2015, **4**, pp. 67 – 70, Novosibirsk, Russia (2015)

4. Soil Adjusted Vegetation Index // Index Database. URL: http://www.indexdatabase.de/db/i-single.php?id=87

5. N. Sokolova, D. Sokolov, *The use of vegetation indexes to assess the soil-ecological condition of technogenic landscapes,* Pochvy v Biosfere: Sbornik Materialov Nauchnoj Konferencii s Mezhdunarodnym Uchastiem, Posvyashchennoj 50-letiyu Instituta Pochvovedeniya i Agrohimii SO RAN [Soils in the Biosphere: Proceedings of the Scientific Conference with International Participation, Dedicated to the 50th Anniversary of the Institute of Soil Science and Agrochemistry SB RAS], 10 September – 14 October 2018, pp. 345 – 349, Tomsk, Russia (2018)

6. S. Drozdov, S. Sladkopevtsev, *Distancionnye metody ocenki prirodnyh resursov (rel'ef i pochvy) [Remote sensing methods for natural resource assessment (landforms and soils)],* 178 p., Moscow, Russia (2015)

AUTHOR INDEX